MAGNETIC ALCHEMY

Magnetic Alchemy

THE CHEMISTRY BEHIND IRON'S ATTRACTION

Rayan Musk

Bliss Enterprise

Contents

INDEX

INTRODUCTION

1. Inroductin
2. Presentation

   Attraction, a crucial power of nature, has charmed and dazzled human interest for quite a long time. From old stories of lodestones directing mariners across the oceans to current innovative wonders like attractive reverberation imaging (X-ray), the narrative of attraction unfurls as an intriguing verifiable odyssey. This outline investigates the quintessence of attraction, following its authentic importance and its extraordinary excursion through different societies, logical disclosures, and innovative progressions.

3. **The Pith of Attraction**
1. **Basics of Attraction**

   At its center, attraction is a power that draws in or repulses objects with attractive properties. Objects that show attractive way of behaving are called magnets, and they can impact other attractive materials without actual contact. The groundwork of attraction lies in the arrangement and development of charged particles, especially electrons, inside iotas.

2. **Attractive Fields: Undetectable Powers**

   Attractive fields, the undetectable locales around magnets, apply impact on adjacent items. These fields reach out into space, making a three-layered zone where attractive powers act. The strength and heading of attractive fields are central properties that administer the way of behaving of magnets.

3. **Attractive Poles: North and South**

Each magnet has two particular poles - a north pole and a south pole. Like shafts repulse one another, while inverse posts draw in. The actual Earth goes about as a goliath magnet, with its attractive north pole close to the geographic North Pole and its attractive south pole close to the geographic South Pole.

III. Early Experiences with Attraction

1. **Antiquated China: The Compass and Lodestones**

   One of the earliest recorded experiences with attraction comes from anti-quated China. Chinese pilots in the Han Tradition (206 BCE - 220 CE) found that specific stones, later recognized as lodestones, had a characteristic attractive property. These lodestones were made into compasses, altering route by lining up with the World's attractive field and empowering exact course finding adrift.

2. **Greek Persona: Attractive Rocks and Philosophical Miracle**

   In old Greece, rationalists, for example, Thales of Miletus noticed the particular idea of specific rocks, later distinguished as lodestones. These stones, when suspended, would in general conform to the north-south heading. The Greeks wondered about this peculiarity however didn't completely fathom the fundamental standards of attraction.

3. **Bedouin Commitments: Attractive Revelations in the Medieval times**

During the Islamic Brilliant Age, researchers like Al-Razi and Al-Kindi investigated the properties of magnets. Their compositions laid the basis for later European comprehension of attraction. The compass, at first acquainted with the Islamic world from China, turned into a critical navigational device during this period.

IV. The Renaissance and the Logical Upset

1. **Gilbert's Showstopper: "De Magnete" (1600)**

   The defining moment in the logical comprehension of attraction accompanied crafted by William Gilbert, an English doctor and normal rationalist. In his fundamental work "De Magnete," distributed in 1600, Gilbert efficiently concentrated on attraction and power. He presented the expression "power" and recognized attractive and non-attractive substances.

2. **Galileo and the Attractive Compass**

   Galileo Galilei, the Italian polymath, perceived the capability of the attractive compass for route. His tests and perceptions added to the comprehension of attractive powers and their functional applications.

3. **Attractive Declination: Edmond Halley's Commitment**

Edmond Halley, most popular for anticipating the arrival of Halley's Comet, made critical commitments to the comprehension of attractive declination — the variety of the World's attractive field at various areas. His work laid the preparation for exact route utilizing compasses.

V. The Time of Edification and Attractive Hypotheses

1. **Attractive Fields and Lines of Power: Hans Christian Ørsted (1820)**

The nineteenth century saw critical advancements in the field of attraction. Danish physicist Hans Christian Ørsted's disclosure of the association among power and attraction denoted a pivotal second. In 1820, Ørsted saw that an electric flow delivers an attractive field, establishing the groundwork for the electromagnetic relationship.

2. **Ampère's Electrodynamics and the Introduction of Electromagnetism**

André-Marie Ampère developed Ørsted's work by figuring out numerical conditions portraying the connection between electric flows and attractive fields. Ampère's commitments to electrodynamics set up for the unification of power and attraction.

VI. The Unification of Power and Attraction

1. **Faraday's Acceptance and Maxwell's Conditions**

Michael Faraday, an English researcher, further high level the comprehension of electromagnetism through his trials on electromagnetic acceptance. Faraday's work laid the basis for James Representative Maxwell, who planned a bunch of conditions — Maxwell's conditions — binding together power and attraction. This stupendous accomplishment prepared for the improvement of electromagnetic hypothesis.

2. **Electromagnetic Waves and Light: Maxwell's Inheritance**

Maxwell's conditions brought together power and attraction as well as anticipated the presence of electromagnetic waves, including light. This significant understanding exhibited the interconnected idea of central powers and essentially affected the fields of optics and broadcast communications.

VII. Attraction in the Cutting edge Period

1. **Mechanical Wonders: Attractive Applications in the twentieth Hundred years**

The twentieth century saw a blast of innovative headways saddling the standards of attraction. The improvement of electric generators, transformers, attractive capacity gadgets, and attractive reverberation imaging (X-ray) reformed ventures and medical services, exhibiting the unavoidable impact of attraction in present day life.

2. **Quantum Mechanics and Attraction: The Infinitesimal Domain**

In the domain of quantum mechanics, researchers investigated the minuscule starting points of attraction. The arrangement of electron turns inside materials, especially those containing attractive components like iron, led to the comprehension of ferromagnetism and other attractive peculiarities at the nuclear level.

1. Brief overview of magnetism and its historical significance

Attraction, a key power of nature, has charmed mankind for a really long time. From the baffling lodestone of old times to the multifaceted comprehension of electromagnetism in the advanced period, the excursion of unwinding the mysteries of attraction has been both spellbinding and extraordinary. This article means to give an extensive outline of attraction, investigating its verifiable importance and the development of how we might interpret this interesting power.

1. **The Antiquated Secrets of Lodestone:**

   The historical backdrop of attraction can be followed back to old times, where the revelation of lodestone, a normally polarized mineral, denoted the start of humankind's interest with attractive powers. Antiquated Greek rationalists, for example, Thales and Aristotle saw that lodestone had the curious property of drawing in iron. The Chinese likewise reported the utilization of lodestone in route around the fourth century BCE, utilizing it as a compass to track down bearing.

   The perplexing properties of lodestone started interest and different speculations were proposed to make sense of its attractive nature. Some accepted that lodestone had a living soul, while others thought it had an association with the stars. Regardless of the absence of a logical comprehension, the utilization of lodestone in route established the groundwork for future revelations in the field of attraction.

2. **The Compass and Early Route:**

   One of the main uses of attraction in history is the advancement of the compass. By suspending a charged needle on a turn, early pilots could decide the World's attractive north and explore with more prominent accuracy. The compass altered sea investigation, empowering mariners to navigate immense seas with expanded certainty.

   The Chinese, who previously used the compass, maintained the innovation a carefully hidden mystery for quite a long time. At last, it spread to the Middle Easterner world and Europe, assuming a significant part in the Period of Investigation. The compass worked with exchange and social trade as well as extended how we might interpret Earth's attractive field.

3. **The Rise of Logical Request:**

   The logical insurgency of the seventeenth century saw a shift from otherworldly clarifications to efficient examinations concerning the idea of attraction. William Gilbert, an English doctor, is in many cases credited as the dad of present day attraction. In his original work, "De Magnete" (1600), Gilbert led broad examinations and laid the foundation for figuring out the Earth as a goliath magnet with attractive posts.

Gilbert's work set up for future turns of events, moving researchers like Robert Boyle and Isaac Newton to investigate the idea of attractive powers. Newton, specifically, made huge commitments by figuring out numerical regulations administering the way of behaving of magnets, laying the basis for the logical technique applied to the investigation of attraction.

4. **Power and Attraction Divulged:**

The eighteenth and nineteenth hundreds of years saw a conjunction of improvements in power and attraction. Hans Christian Ørsted's notable disclosure in 1820 showed the cozy association among power and attraction. Ørsted saw that an electric flow makes an attractive field, laying out a connection between these apparently particular peculiarities.

This disclosure prepared for additional examinations by André-Marie Ampère, Michael Faraday, and James Agent Maxwell. Faraday's examinations with electromagnetic enlistment uncovered the rule behind electric generators and transformers, showing the equal connection among power and attraction. Maxwell's conditions, figured out during the nineteenth hundred years, gave a bound together structure to understanding electromagnetism, bringing together the regulations overseeing power and attraction into a solitary, rich hypothesis.

5. **Electromagnetism and Mechanical Progressions:**

The combination of power and attraction enhanced our hypothetical comprehension as well as prompted extraordinary mechanical headways. The improvement of the message, for example, depended on the standards of electromagnetism to send electrical signs over significant distances. This established the groundwork for worldwide correspondence organizations and upset how data was dispersed.

The nineteenth century additionally saw the development of electromagnetic gadgets, for example, engines and generators, which laid the preparation for the charge of society. Crafted by creators like Michael Faraday and Nikola Tesla assumed a pivotal part in the improvement of these advancements, making way for the electrical upset that would control the twentieth 100 years.

6. **Attraction in the twentieth Hundred years:**

The twentieth century saw phenomenal headways in how we might interpret attraction, driven by both hypothetical leap forwards and mechanical advancements. The improvement of quantum mechanics gave a more profound knowledge into the minuscule universe of particles and their attractive properties. Researchers like Paul Dirac and Werner Heisenberg added to the quantum hypothesis of attraction, making sense of peculiarities like ferromagnetism and antiferromagnetism.

Mechanical utilizations of attraction additionally multiplied in the twentieth 100 years. Attractive reverberation imaging (X-ray), a clinical imaging method in light of atomic attractive reverberation, changed symptomatic medication. Attractive materials tracked down applications in different ventures, from information capacity in attractive hard drives to attractive levitation in high velocity trains.

7. **Contemporary Advances and Future Possibilities:**

In the 21st 100 years, attraction keeps on being a lively field of examination with applications traversing different spaces. Nanotechnology and spintronics, which investigate the control of electron turns for data handling, address state of the art areas of attraction research. The journey for novel attractive materials with interesting properties drives progressing examinations, with likely ramifications for energy capacity, figuring, and then some.

Besides, the investigation of attraction reaches out past Earth. Space investigation missions have uncovered the attractive fields of divine bodies, giving experiences into the crucial cycles administering the universe. Grasping the attractive properties of planets, stars, and universes adds to our more extensive cognizance of astronomy.

B. The mysterious allure of magnets and the fascination with iron's magnetic properties

Magnets have an enamoring persona that has interested humankind for a really long time. From the old charm of lodestone to the advanced wonders of electromagnetism, the connection among magnets and iron has fascinated researchers, wayfarers, and scholars the same. This article investigates the baffling appeal of magnets, digging into the verifiable importance, logical investigation, and social effect of iron's attractive properties.

1. **Old Charm with Lodestone:**

The excursion into the puzzling universe of magnets starts with the revelation of lodestone, a normally happening attractive mineral. In old times, the impossible to miss properties of lodestone, for example, its capacity to draw in iron, ignited wonder and interest. The Greeks, Chinese, and Egyptians generally perceived the otherworldly characteristics of lodestone, crediting it to divine powers or mysterious energies.

The Chinese, around 200 BCE, were among quick to use lodestone for route, fostering the compass and upsetting sea investigation. The puzzling draw of lodestone on iron appeared to rise above the actual domain, making an extraordinary association that dazzled the creative mind of old societies.

2. **The Introduction of Logical Request:**

   The change from otherworldly clarifications to logical request denoted a huge defining moment in the comprehension of magnets and their connection with iron. In the seventeenth hundred years, William Gilbert's weighty work, "De Magnete," established the groundwork for the efficient investigation of attraction. Gilbert's perceptions and examinations uncovered that the actual Earth acted like a goliath magnet with particular attractive shafts.

   As logical request advanced, the center moved from the otherworldly to the experimental. The appeal of magnets became interwoven with the scholarly fervor of revealing the regulations administering their way of behaving. The attractive needle, at first a device for route, turned into a logical instrument, helping specialists in their mission to grasp the key idea of attraction.

3. **Iron's Attractive Character:**

   The interest with iron's attractive properties is well established in its special nuclear design. Iron, in its regular state, is made out of particles with attractive minutes that will generally adjust haphazardly, bringing about an absence of by and large attractive way of behaving. Be that as it may, when presented to an attractive field, the attractive snapshots of iron molecules adjust, making major areas of strength for a power inside the material.

   This inborn capacity of iron to become polarized and show attractive properties separates it from different components. The flexibility of iron, combined with its attractive responsiveness, made it a vital participant in the improvement of attractive devices and instruments since forever ago.

4. **The Compass: An Instrument of Investigation:**

   The compass, a gadget that tackles the attractive properties of lodestone or falsely polarized needles, addresses one of the main uses of iron's attractive charm. Guides and pioneers since the beginning of time depended on the compass to cross tremendous seas and graph new domains.

   The compass worked with route as well as filled the Period of Investigation, empowering mariners to wander into unknown waters with recently discovered certainty. This dependence on the attractive properties of iron essentially changed the course of mankind's set of experiences, interfacing far off societies and reshaping the world guide.

5. **Power and Attraction: A Unique Couple:**

   The nineteenth century saw a significant combination of power and attraction, uncovering a more profound and more multifaceted connection between the two peculiarities. The examinations of researchers like Hans Christian Ørsted, André-Marie Ampère, and Michael Faraday showed the interchange between electric flows and attractive fields, leading to the field of electromagnetism.

   Iron's attractive properties assumed a vital part in this change in perspective.

Electromagnets, made by wrapping an iron center with a loop of wire conveying an electric flow, became necessary parts of different innovative headways. This improvement made ready for the creation of electric engines, generators, and transformers, establishing the groundwork for the charge of society.

6. **Mechanical Wonders: Magnets in the Cutting edge Period:**

The twentieth century saw the inescapable incorporation of magnets into present day innovation, further intensifying their puzzling charm. Attractive materials became fundamental parts in ordinary things, from cooler magnets to the attractive strips on Visas. The improvement of attractive reverberation imaging (X-ray), a clinical imaging method that depends on the attractive properties of nuclear cores, changed medical services diagnostics.

Additionally, the attractive properties of iron tracked down applications in information capacity. Attractive hard drives, usually utilized in PCs, store data by controlling the attractive direction of minuscule iron-put together spaces with respect to a circle. This mechanical jump in information capacity limit and speed has turned into a basic piece of the computerized age.

7. **Iron in the Quantum Domain:**

As logical comprehension progressed into the twentieth hundred years and then some, analysts dug into the quantum domain to unwind the secrets of attraction at the nuclear and subatomic levels. Quantum mechanics uncovered the multifaceted dance of electrons inside materials like iron, making sense of peculiarities like ferromagnetism and antiferromagnetism.

In ferromagnetic materials, similar to press, adjoining nuclear attractive minutes suddenly adjust, making a plainly visible charge. This arrangement continues even without an outside attractive field, adding to the lastingness of magnets. Antiferromagnetic materials, then again, show substituting arrangement of attractive minutes, bringing about extraordinary attractive ways of behaving.

8. **Social Imagery and Analogies:**

Past the domains of science and innovation, magnets and iron have tracked down their direction into social imagery and analogies. The attractive draw between sweethearts is frequently figuratively portrayed, accentuating a powerful fascination. In writing and workmanship, magnets act as strong images of association, solidarity, and the undetectable powers that tight spot individuals together.

Iron, with its relationship with strength and solidness, is regularly utilized in social articulations to convey flexibility and faithfulness. The attraction of iron, both strict and allegorical, has become implanted in human language and imagination, rising above the logical domain to bring out further close to home and social implications.

9. **Future Outskirts: Attraction in the 21st 100 years and Then some:**

In the 21st 100 years, the interest with magnets and iron's attractive properties keeps on driving logical investigation and mechanical development. Research in the arising field of spintronics centers around bridling the twist of electrons for data handling, opening additional opportunities for quicker and more energy-proficient electronic gadgets.

Nanotechnology, with its capacity to control materials at the nanoscale, offers invigorating possibilities for the advancement of novel attractive materials with extraordinary properties. These materials might find applications in regions going from medication and energy stockpiling to processing and correspondence.

Also, the investigation of attractive fields in space, from Earth's magnetosphere to far off cosmic systems, adds to how we might interpret the universe. Attractive reverberation spectroscopy in stargazing permits researchers to concentrate on attractive fields in heavenly bodies, revealing insight into the central cycles molding the universe.

C. Purpose of the book: to delve into the chemistry behind iron's magnetism and explore its implications

The reason for this book is to leave on a captivating investigation of the science behind iron's attraction, unwinding the complexities of this peculiarity and diving into its significant ramifications. Iron, an apparently commonplace component, has attractive properties that have enthralled researchers, engineers, and inquisitive personalities over the entire course of time. By unwinding the science of iron's attraction, we plan to not just demystify this dazzling part of materials science yet in addition uncover the broad results and applications that reach out from the tiny to the astronomical.

1. **Underpinnings of Attraction: A Compound Viewpoint:**

   To comprehend the attraction of iron, we should initially set out on an excursion into the domain of science. At the nuclear level, the way of behaving of iron is administered by the course of action of its electrons. Quantum mechanics has uncovered that the arrangement of these electrons brings about the attractive properties saw in iron and other attractive materials.

   This segment of the book will give a primary comprehension of the electronic construction of iron and the quantum mechanical rules that support its attractive way of behaving. Perusers will investigate ideas like twist, attractive minutes, and electron designs, acquiring experiences into the minuscule world that oversees the plainly visible attraction we see in materials.

2. **Ferromagnetism and Then some: Investigating Iron's Attractive Stages:**

   Iron shows ferromagnetism, a peculiarity where adjoining nuclear attractive minutes adjust immediately, bringing about a perceptible charge. In any case, the story doesn't end there. The book will dive into the different attractive periods of iron, including antiferromagnetism and ferrimagnetism, each with its unmistakable qualities and suggestions.

Understanding these attractive stages is critical for getting a handle on the full range of iron's attractive way of behaving. Through striking clarifications and illustrative models, perusers will acquire a far reaching outline of the different manners by which iron communicates with attractive fields, both inside and remotely.

3. **Mechanical Wonders: Tackling Iron's Attraction:**

With a strong groundwork in the science and physical science of iron's attraction, the book will then move concentration to the pragmatic applications have risen up out of this principal understanding. Iron's attractive properties are not bound to the lab; they saturate our regular daily existences in manners we may not actually understand.

Perusers will investigate the job of iron in the improvement of attractive materials that power a heap of advancements. From the attractive strips on charge cards to the hard drives in our PCs, the ramifications of iron's attraction are woven into the texture of our mechanical presence. The part will enlighten the designing wonders that influence iron's attractive ability, forming enterprises and changing the manner in which we live.

4. **Quantum Mechanics Meets Nanotechnology: The Fate of Iron's Attraction:**

As the account unfurls, the book will look into the future, where the crossing point of quantum mechanics and nanotechnology holds commitments of progressive headways. Nanoscale control of materials acquaints another aspect with our capacity to control and design iron's attractive properties.

Perusers will be directed through state of the art examination and improvements, investigating the capability of nanomagnetic materials in fields like medication, energy capacity, and data handling. The book will introduce a dream of how the combination of quantum mechanics and nanotechnology could open remarkable open doors, preparing for the following boondocks in materials science.

5. **Past Earth: Iron's Attractive Impact in Astronomy:**

Iron's attractive impact reaches out past the limits of our planet. In this part, the book will take perusers on a grandiose excursion, investigating the job of iron in the attractive fields of divine bodies. From Earth's magnetosphere to far off worlds, iron assumes an essential part in molding the attractive scenes of the universe.

Perusers will acquire bits of knowledge into the job of iron in stars, planets, and interstellar space, uncovering the secrets of astrophysical attraction. The ramifications of understanding iron's attractive conduct in the universe are significant, offering hints to the basic cycles that administer the universe.

6. **Social and Emblematic Aspects: Iron's Attraction in Human Articulation:**

Attraction, particularly when related with iron, has formed logical and innovative scenes as well as tracked down its direction into human culture, imagery, and articulation. This part of the book will investigate the social meaning of iron's attraction, following its presence in writing, workmanship, and imagery from the beginning of time.

Perusers will find how the strange appeal of magnets and iron has motivated allegories of fascination, solidarity, and versatility in human language and imagination. By investigating the social aspects, the book means to give an all encompassing comprehension of how iron's attraction has saturated human cognizance, rising above its logical starting points to turn into a figurative power in our aggregate creative mind.

Chapter 1

Foundations Of Magnetism

Attraction, with its mysterious charm and certain effect on the normal world, has enthralled human interest for quite a long time. The underpinnings of attraction are well established in verifiable disclosures, from the antiquated interest with lodestones to the cutting edge figuring out grounded in quantum mechanics. This investigation will unwind the many-sided woven artwork of attraction, following its authentic advancement, diving into basic standards, and inspecting the progress to quantum speculations that support our contemporary comprehension.

1. **Old Wonders: Lodestones and the Compass**
1. **Lodestones: Nature's Attractive Gifts**

    The excursion into the groundworks of attraction starts with the disclosure of lodestones, normally charged rocks. Old human advancements, especially the Chinese and Greeks, wondered about lodestones' capacity to draw in iron. This part will investigate early social purposes of lodestones and the attractive compass, featuring their critical job in route and forming authentic shipping lanes.

2. **Compass Route and Sea Investigation**

The development and refinement of the attractive compass changed route, empowering sailors to investigate unfamiliar regions with expanded accuracy. The interaction between lodestones, the World's attractive field, and the compass laid the foundation for worldwide investigation. This part will dig into the verifiable meaning of the compass and its effect on oceanic experiences.

II. The Ascent of Attractive Speculations: Antiquated Greece to the Renaissance

1. **Philosophical Riddles: Early Greek Speculations**

   Antiquated Greek savants, including Thales and Anaximander, examined the idea of magnets and their puzzling properties. This segment will investigate early philosophical speculations encompassing attraction, laying the preparation for the more precise requests that followed.

2. **The Renaissance: Gilbert and the Attractive Way of thinking**

The Renaissance time saw a resurgence of interest in normal way of thinking, and William Gilbert arose as a significant figure in the investigation of attraction. His fundamental work, "De Magnete," established the groundwork for current attractive hypothesis, presenting ideas like attractive posts and the Earth as a goliath magnet. This segment will dig into Gilbert's commitments and the Renaissance's effect on molding attractive way of thinking.

III. Iron and Its Attractive Character

1. **Natural Experiences: Figuring out Iron's Construction**

   To grasp attraction, one must initially figure out the basic properties of iron. This part will investigate the nuclear and sub-atomic design of iron, revealing insight into the variables that add to its attractive character.

2. **The Attractive Dance of Electrons**

   Attraction is essentially attached to the way of behaving of electrons inside iotas. A nitty gritty assessment of electron turn and attractive minutes will disentangle the infinitesimal beginnings of attraction, making way for the investigation of ferromagnetism and other attractive peculiarities.

3. **Ferromagnetism: Arrangement of Attractive Spaces**

The foundation of iron's attraction lies in ferromagnetism. This segment will explain the idea of attractive spaces and their arrangement, giving a more profound comprehension of how materials like iron become super durable magnets.

IV. The Quantum Universe of Attraction

1. **Quantum Mechanics Disclosed**

   As the groundworks of attraction move into the twentieth 100 years, the stage is set for the combination of quantum mechanics into attractive hypothesis. This segment will present the critical standards of quantum mechanics, for example, wave-molecule duality and quantized energy levels, giving the instruments expected to open the secrets of attraction at the quantum level.

2. **Turn and Attractive Minutes: The Tiny View**

   Quantum mechanics presents the idea of electron turn, a quantum property that assumes a focal part in attraction. This segment will dig into the quantum mechanical beginnings of attractive minutes, offering a minuscule perspective on the attractive way of behaving of materials.

3. **Quantum Hypotheses and Their Applications**

Expanding upon the underpinning of quantum mechanics, researchers created quantum hypotheses of attraction to make sense of a scope of peculiarities. This segment will investigate powerful quantum models like the Heisenberg model and the Ising model, featuring their applications in figuring out attractive materials.

V. Attractive Materials Past Iron

1. **The Attractive Range: A Variety of Materials**

   While iron is an attractive heavyweight, various different materials display attractive properties. This segment will give an outline of the attractive range, investigating ferromagnetic, antiferromagnetic, and ferrimagnetic substances, as well as arising materials like attractive polymers.

2. **Mechanical Utilizations of Attractive Materials**

The commonsense ramifications of attractive materials reach out a long ways past simple interest. Electromagnetism frames the foundation of different advances, from attractive capacity gadgets like hard drives to clinical applications, for example, Attractive Reverberation Imaging (X-ray). This segment will investigate the innovative wonders that emerge from how we might interpret attractive materials.

VI. Difficulties and Future Boondocks

1. **Current Difficulties in Attraction**

   Indeed, even with our refined comprehension of attraction, challenges persevere. This part will address flow secrets and obstacles in attraction research, from making sense of fascinating attractive peculiarities for controlling materials at the nanoscale.

2. **State of the art Exploration in Attractive Materials**

   Progressions in materials science and nanotechnology have opened new outskirts in attractive examination. This segment will highlight state of the art studies and advancements, exhibiting how analysts are pushing the limits of what is conceivable in the domain of attraction.

3. **Likely Applications in Quantum Registering and Cutting edge innovations**

As we stand on the cusp of the quantum processing time, this part will investigate the possible utilizations of attraction in quantum advancements. From quantum data handling to attractive qubits, the marriage of attraction and quantum processing holds invigorating conceivable outcomes.

1.1 Early discoveries: lodestones and the compass

The narrative of early revelations in attraction, lodestones, and the improvement of the compass is a captivating excursion through the records of mankind's set of experiences. These revelations molded the universe of route as well as opened up new roads for logical request and innovative progressions. In this investigation, we will dig into the starting points of attraction, the baffling properties of lodestones, and the progressive effect of the compass on route.

The Beginnings of Attraction

"Magnet" tracks down its foundations in Magnesia, a district in old Greece, where lodestones, a normally happening attractive stone, were first found. The peculiarity of attraction, in any case, goes back a lot further. The antiquated Greeks and Chinese both noticed the attractive properties of lodestones around 600 BCE.

Antiquated Chinese Disclosures

In antiquated China, the idea of attraction was related with the Chinese divination framework known as the "I Ching." The attractive properties of lodestones were bridled for fortune-telling, and the directional properties of magnets were perceived as the compass.

The Chinese likewise seen that when a lodestone was suspended and permitted to turn unreservedly, it conformed to the World's attractive field. This early perception laid the foundation for the advancement of the compass.

Lodestones: Nature's Attractive Wonders

Lodestones, made basically out of the mineral magnetite, became objects of interest because of their capacity to draw in iron. This attractive property was a riddle for early social orders, and the secret of lodestones powered logical interest.

Attraction in Days of yore

The Greeks, especially the thinker Thales of Miletus, are credited with probably the earliest recorded perceptions of lodestones. They noticed that lodestones had the ability to draw in iron and, when suspended, adjusted themselves along a north-south pivot.

The Romans, as well, perceived the mysterious properties of lodestones. The Roman creator and naturalist Pliny the Senior recorded the utilization of lodestones in route in his exhaustive work "Naturalis Historia."

Islamic Commitments

During the Islamic Brilliant Age, researchers made huge commitments to the comprehension of attraction. Al-Biruni, an eleventh century Persian researcher, led investigates the attractive properties of lodestones and added to the developing group of information on attraction.

Ibn al-Haytham, another Islamic researcher, composed broadly on optics yet in addition investigated the standards of attraction. His works established the groundwork for later advancements in the field.

The Compass: Exploring the Unexplored world

The development of the compass is quite possibly of the most extraordinary part throughout the entire existence of route. From its modest beginnings as a straightforward attractive needle to its joining into refined navigational instruments, the compass plays had a critical impact in forming human investigation.

Early Compasses

The main compasses were straightforward gadgets, comprising of a lodestone or attractive needle mounted on a drifting base. These early compasses were probable utilized for divination and didn't yet have an immediate application in route.

The Chinese, who are credited with the innovation of the compass, at first involved it for geomancy — an act of divining the World's energies. Over the long haul, nonetheless, the compass tracked down its direction into oceanic route.

Compass in Sea Investigation

The compass' entrance into sea route is firmly connected to the Chinese and their high level cruising innovations. Chinese sailors started involving compasses for route during the Melody Tradition (960-1279 CE). The attractive needle, pointing north, furnished mariners with a dependable reference for bearing, empowering them to wander out from the dark ocean with expanded certainty.

In the thirteenth hundred years, the compass advanced toward Europe through shipping lanes and associations with the Islamic world. European mariners immediately perceived its worth, and the compass turned into an irreplaceable apparatus for nautical.

Progressions in Compass Innovation

As sea investigation progressed, so compassed innovation. Compasses advanced from basic lodestones to additional modern instruments. The expansion of a compass card set apart with cardinal places (north, south, east, west) and later the coordination of the compass into astrolabes and quadrants upgraded route precision.

The compass likewise assumed a critical part in the Period of Disclosure. Adventurers like Christopher Columbus and Ferdinand Magellan depended on compasses to explore across unfamiliar waters, adding to the planning of the well explored parts of the planet and the disclosure of new landmasses.

Logical Comprehension of Attraction

The comprehension of attraction advanced throughout the long term, progressing from supernatural interest to a more logical investigation. Key figures in this movement incorporate William Gilbert, who is in many cases thought about the dad of attraction, and later researchers who created speculations to make sense of the attractive peculiarity.

William Gilbert's Commitments

In the late sixteenth hundred years, William Gilbert, an English doctor and normal savant, directed broad trials on attraction and recorded his discoveries in the fundamental work "De Magnete." Gilbert recommended that the actual Earth was a goliath magnet, lining up with the perceptions of lodestones.

His work established the groundwork for a more efficient and logical way to deal with the investigation of attraction. Gilbert's bits of knowledge prepared for future advancements in the comprehension of Earth's attractive field.

Speculations of Attraction

As the logical strategy acquired conspicuousness, different hypotheses arose to make sense of the idea of attraction. In the eighteenth hundred years, the French researcher Charles-Augustin de Coulomb figured out Coulomb's Regulation, portraying the power between attractive shafts. This regulation became essential in grasping attractive connections.

In the nineteenth hundred years, James Representative Maxwell's conditions gave a complete system to figuring out electromagnetism, bringing together power and attraction as various parts of a similar power. Maxwell's work laid the basis for the advancement of innovations like electric generators and transformers.

Attraction in Present day Innovation

The comprehension of attraction not just extended our insight into the normal world yet additionally prepared for a bunch of mechanical applications. From electric engines to attractive reverberation imaging (X-ray), attraction has turned into an essential piece of current life.

Electromagnetism and Innovation

The association among power and attraction ended up being a progressive forward leap in innovation. The improvement of electromagnets considered the controlled age of attractive fields, prompting the innovation of electric engines and generators.

The message, one of the earliest significant distance specialized gadgets, depended on the standards of electromagnetism. The transmission of electrical transmissions over significant distances through broadcast wires exhibited the viable uses of electromagnetic innovation.

Attractive Reverberation Imaging (X-ray)

In the field of medication, the revelation of atomic attractive reverberation (NMR) during the twentieth century prompted the improvement of attractive reverberation imaging (X-ray). This painless clinical imaging strategy utilizes strong magnets and radiofrequency waves to make point by point pictures of inside body structures, upsetting analytic medication.

X-ray innovation exhibits the extraordinary effect of attraction on assorted fields, from material science and designing to medical care and then some.

1.2 Development of magnetic theories: from ancient Greece to the Renaissance

The investigation of attractive peculiarities has a celebrated history that unfurls across hundreds of years, beginning with the early perceptions in old Greece and arriving at a zenith of understanding during the Renaissance. This excursion, set apart by the persona of lodestones, the innovation of the compass, and the rise of efficient logical request, gives an interesting story of human interest and scholarly development.

Old Greece: Thales and the Mystery of Lodestones

The starting points of attractive hypotheses can be followed back to antiquated Greece, where the thinker Thales of Miletus made the absolute earliest recorded perceptions of attraction around 600 BCE. Thales found that a curious stone, later distinguished as magnetite, had a novel property - the capacity to draw in iron.

Thales' Perceptions

Thales' interest with lodestones was essentially revolved around their puzzling ability to draw in iron articles. He saw that when a piece of iron was brought close to a lodestone, it would be drawn toward it. This straightforward yet significant perception established the groundwork for the comprehension of attraction.

The Greeks, while without a far reaching logical structure, likewise saw that when a lodestone was suspended and permitted to turn openly, it adjusted itself along a north-south hub. This conduct indicated an association between the lodestone and the World's attractive field, albeit the full ramifications of this peculiarity were not yet understood.

Antiquated China: Compass and Divination

While the Greeks were disentangling the secrets of lodestones, an equal improvement was occurring in old China. The Chinese perceived the attractive properties of lodestones as well as cleverly applied them to commonsense purposes.

Chinese Compass and Divination

In antiquated China, lodestones were at first utilized for divination as opposed to route. The Chinese saw that a lodestone, when permitted to pivot uninhibitedly, would get comfortable a north-south arrangement. This directional property of lodestones became instrumental in the advancement of the compass.

The earliest Chinese compasses were utilized in divination rehearses known as geomancy. A spoon-molded lodestone was mounted on a stage, and the heading wherein the lodestone pointed was thought of as significant for divining the World's energies. After some time, this divination device developed into a navigational instrument.

Islamic Brilliant Age: Progressions in Attraction

The light of information passed from old Greece to the Islamic world during the Brilliant Age (eighth to fourteenth hundred years). Islamic researchers based after existing information and directed trials to develop the comprehension of attraction.

Al-Biruni's Tests

One eminent figure from the Islamic Brilliant Age is Al-Biruni, an eleventh century Persian researcher. Al-Biruni led analyses to investigate the attractive properties of lodestones. His work included precise perceptions and estimations, contributing important bits of knowledge into the way of behaving of attractive materials.

Ibn al-Haytham's Experiences

Ibn al-Haytham, known for his commitments to optics, likewise investigated attraction. While his essential shine was on light and vision, his perceptions on attraction added to the developing group of information. These Islamic researchers prepared for the transmission of information to archaic Europe.

Archaic Europe: Guido Guinizelli and the Compass

As information from antiquated Greece and the Islamic Brilliant Age advanced toward archaic Europe, researchers started embracing viable uses of attractive standards.

Guido Guinizelli and the European Compass

In the twelfth hundred years, Guido Guinizelli, an Italian researcher, is attributed with acquainting the compass with Europe. The compass, which had been utilized for divination and route in China, presently tracked down its direction under the control of European mariners and guides.

The European compass, at first a lodestone needle drifting in water, turned into a fundamental device for oceanic route. Mariners could now decide their heading in any event, when away from land, opening up additional opportunities for investigation and exchange.

Renaissance: William Gilbert and the Introduction of Current Attraction

The Renaissance time denoted a time of significant scholarly restoration in Europe, and attraction turned into a subject of methodical logical request.

William Gilbert's "De Magnete"

The critical second in the advancement of attractive speculations accompanied crafted by William Gilbert, an English doctor and normal thinker. In 1600, Gilbert distributed "De Magnete," a notable composition that established the groundwork for present day attraction.

Gilbert dismissed before ideas that magnets got their power from heavenly bodies. All things being equal, he recommended that the actual Earth was a monster magnet. Through fastidious analyses, Gilbert showed that the World's attractive field impacted the way of behaving of lodestones.

Gilbert's Commitments

"De Magnete" covered different parts of attraction, including the properties of lodestones, attractive acceptance, and the way of behaving of magnets. Gilbert presented the idea of attractive poles, recognizing the north-chasing and south-chasing poles of a magnet.

Gilbert's orderly and observational way to deal with concentrating on attraction denoted a takeoff from the speculative and otherworldly thoughts of prior times. His work laid the foundation for future headways in the comprehension of Earth's attractive field.

Hypotheses of Attraction: seventeenth to nineteenth Hundreds of years

Expanding upon Gilbert's establishments, resulting hundreds of years saw the definition of different hypotheses that intended to make sense of the idea of attraction.

Cartesian Attraction

In the seventeenth 100 years, René Descartes proposed a hypothesis known as Cartesian attraction. Descartes recommended that space was loaded up with twirling vortices of small particles, and these vortices caused attractive peculiarities. While the Cartesian model didn't endure for an extremely long period, it addressed an early endeavor to give a robotic clarification to attraction.

Coulomb's Regulation

In the eighteenth hundred years, Charles-Augustin de Coulomb figured out Coulomb's Regulation. This regulation depicted the power between attractive posts and depended on cautious exploratory perceptions. Coulomb's work was a huge step in the right direction in measuring attractive communications.

Ampère's Electrodynamics

In the nineteenth hundred years, André-Marie Ampère added to the comprehension of attraction with his work on electrodynamics. Ampère's investigations uncovered the connection between electric flows and attraction, laying the preparation for the later unification of power and attraction.

Maxwell's Electromagnetic Hypothesis: Unification of Powers

The capstone of the improvement of attractive hypotheses came in the nineteenth 100 years with crafted by James Assistant Maxwell.

Maxwell's Conditions

James Representative Maxwell figured out a bunch of conditions during the nineteenth century that gave a binding together system to grasping power and attraction. Maxwell showed that electric and attractive fields were interconnected and could spread through space as electromagnetic waves.

Maxwell's conditions addressed a great accomplishment in the unification of the powers of power and attraction. This combination established the groundwork for the advancement of innovations like electric generators, transformers, and, eventually, remote correspondence.

An Excursion of Disclosure

The improvement of attractive hypotheses from old Greece to the Renaissance addresses an enamoring excursion of human investigation and scholarly advancement. From the early perceptions of lodestones by Thales to the methodical requests of Gilbert and the bringing together speculations of Maxwell, every time added to our advancing comprehension of attraction.

The change from magic to efficient request reflected the more extensive scholarly changes of the Renaissance. As attractive hypotheses progressed, they made sense of regular peculiarities as well as laid the basis for mechanical advancements that keep on molding our cutting edge world. The narrative of attractive improvement is a demonstration of the voracious interest of the human brain and its capacity to disentangle the secrets of the universe.

1.3 Magnetic materials beyond iron: exploring the magnetic spectrum

While iron has for some time been related with attraction because of its normal attractive properties, the attractive range reaches out a long ways past this solitary component. A different cluster of materials shows attractive way of behaving, going from regular substances to extraordinary mixtures. This investigation digs into the captivating universe of attractive materials past iron, revealing the variety of attractive peculiarities and their applications.

Ferromagnetism: Past Iron

Ferromagnetism, portrayed by the arrangement of attractive minutes in a material even without a trace of an outside attractive field, broadens well past iron. Iron, cobalt, and nickel are the exemplary ferromagnetic materials, however different components and mixtures likewise show this fascinating way of behaving.

Cobalt and Nickel

Cobalt and nickel, arranged close by iron in the occasional table, share comparable ferromagnetic properties. These components, known as the iron group of three, can be charged and hold their attractive state after the expulsion of an outer attractive field. Cobalt and nickel find applications in the assembling of magnets, especially in the development of solid extremely durable magnets utilized in different advances.

Intriguing Earth Magnets

The lanthanide series of components, generally alluded to as interesting earth components, incorporates neodymium and samarium, which are basic parts of strong uncommon earth magnets. These magnets, frequently joined with different components like boron, show remarkable attractive strength, making them significant in applications going from electric engines and earphones to clinical gadgets.

Antiferromagnetism: The Attractive Artful dance

In antiferromagnetic materials, adjoining attractive minutes adjust in inverse bearings, bringing about a crossing out of the generally speaking attractive second. While this counterbalances plainly visible polarization, antiferromagnetic materials are wealthy in tiny attractive associations.

Manganese Oxides

Manganese oxides, especially in specific precious stone designs, exhibit antiferromagnetic way of behaving. In these materials, contiguous manganese particles display contradicting attractive minutes, making an attractive artful dance where the generally speaking attractive impact is invalidated. Understanding antiferromagnetism is critical for applications in spintronics, a field that investigates the utilization of electron turn for data handling and stockpiling.

Ferrimagnetism: The Split the difference

Ferrimagnetic materials, as antiferromagnets, have restricting attractive minutes, however the extents of these minutes are unique. This outcomes in a net attractive second, making ferrimagnets show perceptible charge.

Magnetite (Fe3O4)

Magnetite, otherwise called lodestone, is an exemplary illustration of a ferrimagnetic material. It comprises of iron(II) and iron(III) particles with various attractive minutes, making a net attractive second. Magnetite has been utilized generally for route purposes and keeps on being read up for its extraordinary attractive properties.

Paramagnetism: The Attractive Reaction

Paramagnetic materials show a feeble fascination with an outer attractive field because of the presence of unpaired electrons. While this attractive reaction is for the most part powerless and brief, it is a predominant peculiarity in different materials.

Aluminum and Platinum

Aluminum and platinum are instances of paramagnetic materials. Within the sight of an attractive field, the unpaired electrons in these materials conform to the field, prompting a powerless attractive fascination. While not quite as articulated as ferromagnetism, paramagnetism assumes a part in different fields, including materials science and science.

Diamagnetism: The Loathsome Power

Rather than paramagnetism, diamagnetic materials are repulsed by an outside attractive field. This peculiarity emerges from the prompted attractive second contradicting the applied field, bringing about a powerless shock.

Bismuth and Graphite

Bismuth is a notable diamagnetic material. When presented to an attractive field, bismuth displays an inconspicuous loathsome power, exhibiting the diamagnetic impact. Furthermore, certain types of carbon, like graphite, additionally show diamagnetic properties. Diamagnetism is frequently taken advantage of for levitation tests and attractive safeguarding applications.

Superparamagnetism: Small Magnets at Play

At the nanoscale, certain materials show superparamagnetic conduct, where little attractive particles briefly line up with an outer attractive field.

Attractive Nanoparticles

Attractive nanoparticles, made out of materials like iron oxide, show superparamagnetism. These nanoparticles find applications in medication, especially in attractive reverberation imaging (X-ray) and designated drug conveyance. Their little size permits them to interface with natural frameworks at the cell and sub-atomic levels.

Turn Glasses: Attractive Turmoil

Turn glasses address an extraordinary condition of attractive issue in specific materials, where attractive minutes show a disarranged plan.

Combinations and Mixtures

Combinations and mixtures, frequently made out of change metals, can show turn glass conduct. In turn glasses, the attractive snapshots of the constituent molecules or particles cooperate in a scattered way, prompting a complex attractive state. Understanding twist glasses is pivotal for propelling our insight into attractive stage changes and the more extensive field of consolidated matter physical science.

Applications and Mechanical Advances

The assorted scope of attractive materials past iron has energized mechanical headways across different fields.

Attractive Recording Media

The attractive properties of materials like cobalt and nickel are critical in attractive recording media, for example, hard plate drives. The capacity of these materials to hold polarization takes into account the capacity and recovery of computerized data.

Attractive Reverberation Imaging (X-ray)

In the field of medication, the superparamagnetic conduct of iron oxide nano-particles assumes a urgent part in attractive reverberation imaging (X-ray). These nanoparticles upgrade the differentiation of MR pictures, giving significant demonstrative data in a harmless way.

Magnetostrictive Materials

Magnetostrictive materials, which change shape because of an applied attractive field, are used in sensors and actuators. Terfenol-D, a compound of terbium, dysprosium, and iron, is an outstanding magnetostrictive material utilized in different applications, including sonar frameworks and vibration damping gadgets.

Spintronics

Figuring out the attractive properties of materials, particularly with regards to antiferromagnetism, is urgent for the field of spintronics. Spintronics investigates the control of electron turn for data handling and stockpiling, offering expected headways in figuring and information stockpiling advancements.

Difficulties and Future Headings

While critical headway has been made in understanding and using attractive materials, challenges and unanswered inquiries persevere.

Energy-Proficient Materials

Endeavors are in progress to find and plan materials with worked on attractive properties for energy-proficient applications. This incorporates the quest for materials with upgraded attractive coupling, decreased energy misfortunes, and expanded security.

Quantum Magnets

The investigation of quantum magnets, which show special attractive ways of behaving at the quantum level, addresses an outskirts in research. Quantum magnets could have suggestions for quantum figuring and the improvement of novel materials with fascinating attractive properties.

Economical Attractive Materials

As innovation progresses, there is a developing accentuation on creating supportable attractive materials. This includes investigating options in contrast to uncommon earth components and limiting the ecological effect of attractive advances.

Chapter 2

Iron And Its Magnetic Personality

Iron, with its attractive charm, plays had a urgent impact in molding the course of mankind's set of experiences. From old compasses directing adventurers across the oceans to present day advancements fueling our regular routines, iron's attractive character has made a permanent imprint on science, route, and industry. This investigation digs into the complex connection among iron and attraction, following its authentic importance, disentangling the logical standards at play, and analyzing the different applications that have arisen throughout the long term.

The Authentic Attraction of Iron

The attractive properties of iron have been perceived and tackled by civilizations since forever ago, adding to headways in route, science, and social practices.

Old Compasses and Route

The old Chinese were among quick to saddle the attractive properties of iron for route. Around the fourth century BCE, Chinese guides started utilizing compasses made of charged iron needles to line up with the World's attractive field. This inventive utilization of iron permitted mariners to explore with more prominent accuracy, opening up additional opportunities for investigation and exchange.

In Europe, the reception of the compass was an extraordinary second during the Medieval times. Guides outfitted with iron compasses could wander into unfamiliar waters, depending on the solid arrangement of the needle with the World's attractive field. The compass turned into a crucial instrument for sea investigation, adding to the Time of Disclosure and the planning of the explored parts of the planet.

Social and Custom Importance

Iron's attractive character stretched out past pragmatic applications to social and custom importance in different social orders. In antiquated societies, iron was frequently connected with strength, power, and, surprisingly, enchanted properties.

Iron curios, like weapons and apparatuses, were in some cases permeated with representative implications and utilized in strict functions.

The Logical Underpinnings: Figuring out Iron's Attraction

The attractive character of iron is established in its nuclear and atomic design, administered by the way of behaving of electrons.

Nuclear Design and Attractive Minutes

Iron, with its nuclear number 26, has a clear cut nuclear design. The attractive character of iron emerges from the way of behaving of its electrons, especially those in the furthest shell. In an attractive field, these electrons adjust their attractive minutes, making a generally speaking attractive second for the iron molecule.

Ferromagnetism: The Aggregate Arrangement

Iron is a ferromagnetic material, showing an aggregate arrangement of attractive minutes in a particular heading. At the minuscule level, this arrangement brings about attractive spaces, locales inside the material where the attractive minutes are equal. Without a trace of an outer attractive field, these spaces might point in irregular headings, bringing about a net polarization of nothing.

When presented to an outer attractive field, the attractive spaces in ferromagnetic materials like iron line up with the field, making a plainly visible polarization. This arrangement perseveres even after the expulsion of the outside field, bringing about a super durable magnet.

Curie Temperature: The Attractive Change

The attractive character of iron goes through a huge progress at the Curie temperature. Underneath this basic temperature, iron shows ferromagnetic way of behaving, with adjusted attractive minutes. Over the Curie temperature, warm disturbance upsets the arrangement, and iron loses its ferromagnetic properties.

Understanding the Curie temperature is vital for applications including temperature-delicate magnets. Amalgams and mixtures can be designed to display explicit attractive ways of behaving at wanted temperatures, growing the scope of utilizations in different conditions.

Uses of Iron's Attractive Character

Iron's attractive character has tracked down boundless applications across different fields, from innovation and industry to medical care and then some.

Electromagnetic Gadgets

Iron is a basic part in the development of electromagnets. At the point when an electric flow courses through a loop twisted around an iron center, the attractive snapshots of the iron iotas adjust, making serious areas of strength for a field. Electromagnets are indispensable to different gadgets, including electric engines, transformers, and attractive reverberation imaging (X-ray) machines.

Electric Engines

The attractive character of iron is tackled in electric engines, where the association between attractive fields and electric flows produces rotational movement.

Iron centers as overlaid sheets are much of the time used to upgrade the attractive properties of engine parts, guaranteeing productive energy change.

Transformers

Transformers, fundamental for power conveyance, use iron centers to work with the effective exchange of electrical energy between curls. The attractive character of iron empowers the enlistment of an attractive field, prompting voltage change in the optional curl.

Attractive Reverberation Imaging (X-ray)

In the field of medical care, iron's attractive properties are utilized in attractive reverberation imaging (X-ray). Iron oxide nanoparticles, with their superparamagnetic conduct, improve the difference in X-ray checks, giving itemized pictures of inner designs. This application exhibits the adaptability of iron's attractive character in propelling clinical diagnostics.

Attractive Capacity Media

Iron's attractive character is at the center of attractive stockpiling media, for example, hard plate drives (HDDs). In these gadgets, changes in the attractive direction of iron-based materials encode and recover advanced data. The accuracy and solidness of iron's attractive properties add to the dependability of information stockpiling advances.

Magnetite in Route

Magnetite, a normally happening iron oxide, assumed a noteworthy part in route. Lodestones, made basically out of magnetite, were utilized as early compasses by antiquated sailors. The attractive character of magnetite permitted mariners to explore by adjusting the lodestone to the World's attractive field.

Iron in Social Curios

Iron's attractive character has been protected in social antiques, from old weapons to imaginative manifestations. Iron ancient rarities frequently convey representative and verifiable importance, mirroring the getting through effect of iron on human progress.

Difficulties and Advancements in Iron Attraction

While iron's attractive character has been widely used, continuous examination looks to address difficulties and investigate imaginative applications.

Uncommon Earth Choices

The dependence on uncommon earth components, for example, neodymium and samarium, in the development of strong long-lasting magnets presents difficulties because of their shortage and natural effect. Specialists are investigating elective materials, including iron-based compounds and combinations, to create feasible and eco-accommodating magnets with practically identical execution.

Nanomaterials and Spintronics

Propels in nanotechnology have prompted the investigation of iron-based nanomaterials for applications in spintronics. Spintronics use the twist of electrons for data handling and stockpiling, opening up opportunities for additional energy-productive and superior execution electronic gadgets.

Magnetostrictive Advancements

Research in magnetostrictive materials, including specific iron combinations, plans to upgrade their presentation for applications in sensors and actuators. Advancements in magnetostriction innovation can possibly work on the productivity of gadgets in fields, for example, mechanical technology and primary wellbeing observing.

Iron and What's in store: An Attractive Odyssey

As we explore the future, the attractive character of iron keeps on spellbinding analysts, designers, and trailblazers. From the minuscule arrangement of attractive minutes to the plainly visible applications in innovation, iron's attractive properties have left a permanent engraving on the embroidered artwork of logical revelation and mechanical headway.

The continuous investigation of iron's attractive character holds the commitment of practical arrangements, novel applications, and a more profound comprehension of the hidden standards overseeing attractive way of behaving. As we leave on this attractive odyssey, the persevering through connection among iron and attraction stays a reference point directing us toward new outskirts in science and innovation.

2.1Elemental properties of iron

Iron, a natural foundation of the occasional table, holds a particular spot in the story of mankind's set of experiences and mechanical development. With its nuclear number 26 and image Fe, iron shows a mind boggling snare of essential properties that have formed its importance across different spaces. This extensive investigation plunges into the nuclear construction, physical and synthetic ascribes, overflow, authentic pertinence, organic ramifications, and ecological contemplations connected with iron.

Nuclear Construction of Iron

At the core of iron's essential personality lies its nuclear design. Situated in Gathering 8 and Period 4 of the occasional table, iron shows a mind boggling plan of 26 protons and 26 neutrons inside its core. The electron arrangement [Ar] 3d^6 4s^2 means the appropriation of electrons in shells, uncovering iron's synthetic way of behaving and reactivity.

Progress metals, a classification to which iron has a place, are described by special properties like variable oxidation states and the capacity to shape complex particles. On account of iron, oxidation states +2 and +3 prevail, with the +2 state being all the more generally experienced.

Actual Properties of Iron

Thickness and Hardness

Iron is prestigious for its thickness, tipping the scales at roughly 7.87 grams per cubic centimeter. This trademark, combined with its intrinsic hardness, originates from its glasslike structure. The game plan of molecules in an efficient cross section gives strength and unbending nature to press.

Softening and Edges of boiling over

Iron endures impressive temperatures. With a softening mark of 1,538 degrees Celsius (2,800 degrees Fahrenheit) and a limit of 2,861 degrees Celsius (5,182 degrees Fahrenheit), iron grandstands dependability and versatility, making it essential in different high-temperature applications.

Attractive Properties

Iron's attractive ability is a particular component. It has solid attractive properties, permitting it to be polarized fundamentally. This trait has prepared for its use in the creation of magnets and its joining into assorted electronic gadgets.

Variety and Gloss

In its unadulterated state, iron flaunts a silver-dark shade and a metallic brilliance. Notwithstanding, openness to dampness and oxygen sets off a change, prompting the development of a ruddy earthy colored layer of iron oxide, ordinarily perceived as rust.

Compound Properties of Iron

Reactivity

Iron's reactivity is moderate. While lifeless with water at room temperature, iron promptly draws in with acids, yielding hydrogen gas. The liking for oxygen within the sight of dampness brings about the development of iron oxide or rust, a urgent cycle adding to the consumption of iron-based structures.

Oxidation States

The capacity of iron to exist in different oxidation states is a demonstration of its compound flexibility. Iron dominatingly takes on the +2 and +3 oxidation states, assuming a pivotal part in redox responses. These changes are crucial in both natural capabilities and modern applications.

Complex Arrangement

Iron's ability to shape complex particles and mixtures emerges from its variable oxidation states. These edifices, frequently showing energetic varieties and exceptional properties, contribute essentially to the domain of coordination science.

Instability

While iron itself isn't burnable, finely separated iron particles or press residue can represent a fire danger. This trademark is especially pertinent in businesses where iron powders are used, requesting cautious dealing with and capacity rehearses.

Overflow and Event

Iron stands as quite possibly of the most plentiful component on The planet, comprising roughly 5% of the World's hull. The essential supplies of iron are tracked down as iron metals, with hematite (Fe_2O_3) and magnetite (Fe_3O_4) driving the program. The mining and ensuing handling of these minerals work with the extraction of iron, which turns into a fundamental material in various businesses.

Significant iron stores are moved in nations like Australia, Brazil, China, India, and Russia. The boundless dispersion of iron metal has generally delivered it a financially savvy and pervasively utilized material.

Authentic Importance and Advancement

Iron Age

The verifiable direction of iron is entwined with the advancement of human civilization. The coming of iron utilization denoted an essential age known as the Iron Age. Changing from the Bronze Age, social orders all over the planet embraced the groundbreaking capability of iron, producing instruments and weapons that moved mechanical headways.

Iron's authority during the Iron Age was filled by its better properties thought about than different metals. Iron devices, prestigious for their toughness and strength, became instrumental in agribusiness, development, and fighting, reshaping cultural designs and exchange elements.

Metallurgical Advancements

The development of iron handling methods, from bloomery heaters to impact heaters, mirrors the inventiveness of metallurgists across various societies and ages. Developments like the impact heater reformed iron creation, empowering bigger scope extraction and refining processes that filled the modern upheaval.

Modern Unrest

The Modern Unrest denoted a seismic change in human efficiency, and iron assumed a focal part. The advancement of steam motors, rail routes, and apparatus depended on the accessibility of iron and its combinations, especially steel. The adaptability of iron in assembling and development slung social orders into a period of remarkable mechanical progression.

Contemporary Applications

Metallurgy in the Advanced Period

The tradition of iron keeps on resounding in present day metallurgy. Iron and its combinations, strikingly steel, are pervasive in development, transportation, and assembling. The improvement of particular combinations has extended the application range, yielding materials customized to explicit enterprises and innovative necessities.

Development and Foundation

Iron's power and pliability make it a fundamental material in the development business. The development of structures, spans, and other foundation projects depends intensely on iron and steel. Built up concrete, a composite material consolidating steel bars, epitomizes the cooperative energy among iron and different materials in accomplishing primary respectability.

Transportation

Iron's effect on transportation is significant. Steel, a combination overwhelmingly made out of iron, is a vital participant in the car, aviation, and oceanic enterprises.

The strength and solidness of iron and its composites add to the assembling of vehicles, boats, trains, and airplane, molding the scene of current transportation.

Hardware and Gear

The hardware and gear in assorted enterprises owe their reality to the adaptability of iron. Gears, motors, pipelines, and a variety of mechanical parts are made from iron and its compounds. The material's capacity to endure anxiety makes it key in the hardware that powers economies.

Attractive Applications

Iron's attractive properties track down pragmatic applications in various mechanical spaces. From electric engines to generators and electronic gadgets, iron's job in making magnets and attractive materials is instrumental. Attractive capacity media, for example, hard drives, influence iron-based materials for information capacity and recovery.

Organic Significance

Past its modern and innovative importance, iron is complicatedly woven into the texture of natural frameworks. In the human body, iron is a principal part of hemoglobin, the protein liable for oxygen transport in the blood. Iron's part in enzymatic responses and cell processes highlights its imperativeness forever.

Challenges and Natural Effect

While iron has been a sturdy friend in human advancement, its extraction and usage present significant difficulties and ecological results.

Mining and Ecological Effect

The extraction of iron mineral includes critical natural effects. Open-pit mining and the utilization of large equipment upset biological systems, prompting territory annihilation and soil disintegration. Besides, the removal of mining waste, frequently loaded down with pollutants, represents a danger to water quality in neighboring waterways and streams.

Erosion and Upkeep

The erosion of iron designs, bringing about the development of rust, is an unending test. The financial weight of keeping up with and it is significant to fix consumed framework. High level covering advancements and erosion safe compounds plan to alleviate this test, yet carefulness and proactive upkeep stay fundamental.

Energy Power

Iron and steel creation are energy-concentrated processes, adding to ozone depleting substance discharges. Customary strategies, for example, the impact heater process, depend on carbon-concentrated energizes, worsening natural worries. Continuous endeavors center around creating cleaner and more supportable strategies, for example, hydrogen-based iron creation, to lessen the business' carbon impression.

Squander The board

The development of iron creates huge side-effects and waste, including slag and residue. Legitimate waste administration techniques are fundamental to limit ecological effect. Reusing scrap iron and steel is a pivotal part of practical waste administration, preserving assets and lessening the requirement for extra extraction.

Economical Practices and Future Possibilities

Reusing Drives

As the world wrestles with natural difficulties, reusing arises as an intense technique to moderate the effect of iron extraction. Reusing scrap iron and steel monitors assets as well as decreases energy utilization and ozone harming substance discharges. The round economy model empowers the coordination of reused materials into the creation cycle, advancing maintainability.

Cleaner Innovations

The journey for cleaner and more practical iron creation innovations is picking up speed. Advancements, for example, direct decrease strategies utilizing hydrogen or sustainable power sources expect to decouple iron creation from carbon-escalated processes. These innovations offer the possibility to change the iron and steel industry, adjusting it to worldwide manageability objectives.

Roundabout Economy and Upcycling

Embracing a round economy approach includes planning items and cycles that limit squander and empower the reuse of materials. Upcycling, which includes reusing waste materials into higher-esteem items, presents valuable open doors to reconsider the existence pattern of iron-based items. This change in perspective could rethink the connection between utilization, creation, and ecological effect.

2.2 The role of electrons in magnetism

Attraction is an essential power of nature that has interested researchers for quite a long time. The peculiarity of attraction is intently attached to the way of behaving of electrons inside materials. Electrons, as charged particles, assume a significant part in the appearance of attractive properties in substances. Understanding the connection among electrons and attraction is pivotal for hypothetical material science as well as for the advancement of different innovative applications, for example, attractive capacity gadgets and attractive reverberation imaging (X-ray) machines. In this investigation, we will dig into the mind boggling association among electrons and attraction, investigating the quantum mechanical rules that oversee their way of behaving and their job in the creation and control of attractive fields.

Essential Ideas of Attraction:

To appreciate the job of electrons in attraction, accepting a few fundamental ideas of magnetism is fundamental. At the naturally visible level, attraction is frequently connected with the arrangement of attractive minutes inside a material. Attractive minutes emerge from the twist and orbital movement of electrons. The twist of an electron is a natural property, and it acts like a minuscule attractive dipole.

Moreover, the orbital movement of electrons around the core additionally adds to the generally attractive snapshot of a particle.

Quantum Mechanics and Electron Twist:

In the domain of quantum mechanics, the way of behaving of electrons is depicted by wave capabilities and likelihood circulations. The Pauli Rejection Standard expresses that no two electrons in an iota can have a similar arrangement of quantum numbers, including turn. This prompts the idea of electron turn, which can be by the same token "up" or "down." The twist of electrons is a quantum property that leads to their attractive second.

The Harsh Gerlach analyze, directed in 1922, gave trial proof to the quantization of electron turn. The outcomes showed that electrons have an inborn rakish force, or twist, and that this twist is quantized along a particular heading. This quantization of twist has significant ramifications for the attractive way of behaving of materials.

Ferromagnetism:

Ferromagnetism is quite possibly of the most notable attractive peculiarity, and it is portrayed by the unconstrained arrangement of attractive minutes inside a material. This arrangement prompts the production of plainly visible attractive areas. In ferromagnetic materials, for example, iron and cobalt, adjoining attractive minutes will quite often adjust lined up with one another, subsequent in a solid generally speaking attractive field.

The way to ferromagnetism lies in the communication between electron turns. At the point when nearby electrons have equal twists, their attractive minutes build up one another, prompting a more grounded by and large attractive impact. In ferromagnetic materials, the trade cooperation, a quantum-mechanical impact emerging from the cross-over of electron wave capabilities, is liable for adjusting adjoining turns in a similar bearing.

Antiferromagnetism and Ferrimagnetism:

While ferromagnetism is described by the arrangement of twists in a similar bearing, antiferromagnetism includes the arrangement of twists in inverse headings. In antiferromagnetic materials, adjoining attractive minutes will more often than not adjust antiparallel to one another. This prompts a retraction of the in general attractive second at the perceptible level.

Ferrimagnetism is a more complicated type of attraction where two sublattices with various attractive minutes are available. The vital distinction among ferrimagnetism and antiferromagnetism is that the attractive snapshots of the two sublattices in ferrimagnetic materials are inconsistent, bringing about a net attractive second.

In both antiferromagnetism and ferrimagnetism, the way of behaving of electrons and their twist directions assume an essential part in deciding the attractive properties of the material. The trade connection likewise assumes a huge part in these attractive peculiarities.

Paramagnetism and Diamagnetism:

Paramagnetism and diamagnetism are two other attractive peculiarities that emerge from the way of behaving of electrons in materials. Paramagnetic materials have unpaired electrons, prompting a net attractive second. Within the sight of an outside attractive field, the attractive snapshots of individual molecules or particles in paramagnetic materials will generally line up with the field, bringing about a feeble, positive powerlessness.

Diamagnetic materials, then again, have all their electron turns matched, bringing about a net attractive snapshot of no without an outer attractive field. Nonetheless, when exposed to an outer attractive field, diamagnetic materials foster a frail, negative vulnerability as their electron circles experience a diamagnetic reaction.

Electron Attractive Reverberation:

Electron attractive reverberation (EMR), otherwise called electron paramagnetic reverberation (EPR), is a strong spectroscopic method that takes advantage of the attractive properties of electrons. EMR is especially valuable in concentrating on unpaired electrons in paramagnetic materials. In this strategy, an example is presented to an attractive field, and microwave radiation is applied. The reverberation condition happens when the energy of the microwave matches the energy distinction between two electron turn states.

EMR has applications in different fields, including science, science, and physical science. It is broadly used to concentrate on the electronic design of revolutionaries, abandons in precious stones, and metal particles in natural frameworks. The capacity to test unpaired electron turns gives significant bits of knowledge into the nearby climate and collaborations of electrons inside a material.

Attractive Areas and Hysteresis:

At the perceptible level, attractive materials display the development of attractive spaces. Every space comprises of a district where the attractive snapshots of iotas or particles are adjusted in a specific course. The direction of these attractive areas decides the generally attractive properties of the material. Without any an outer attractive field, the attractive spaces inside a material might show irregular directions, bringing about a net attractive snapshot of nothing.

At the point when an outside attractive field is applied, the attractive areas will quite often line up with the field. As the strength of the outside field expands, an ever increasing number of spaces adjust, prompting a general expansion in the material's charge. This peculiarity is known as attractive immersion.

Hysteresis is a trademark conduct of attractive materials where the charge of the material lingers behind changes in the applied attractive field. During the time spent charge and demagnetization, hysteresis circles portray the connection between the attractive field and the polarization of the material. The state of the hysteresis circle is impacted by different variables, including the attractive anisotropy and the communications between adjoining attractive minutes.

Quantum Mechanical Model of Attraction:

To comprehend the job of electrons in attraction at a more principal level, the quantum mechanical model of attraction gives significant experiences. In this model, the trade connection between electrons is portrayed utilizing quantum mechanics. The trade communication is a quantum-mechanical impact that emerges from the lack of definition of electrons.

As per the Pauli Prohibition Guideline, no two electrons in an iota can possess a similar quantum state. At the point when electrons have equal twists, the trade communication is ideal, prompting a lower energy state. Conversely, when electrons have antiparallel twists, the trade communication is ominous, bringing about a higher energy state.

The trade communication is a vital consider deciding the attractive properties of materials. In ferromagnetic materials, the trade collaboration favors equal arrangement of twists, prompting the development of attractive spaces with a net plainly visible charge. In antiferromagnetic materials, the trade collaboration favors antiparallel arrangement of twists, bringing about a scratch-off of the generally attractive second.

Attractive Anisotropy:

Attractive anisotropy is a peculiarity where the attractive properties of a material rely upon the bearing of the attractive field. It is impacted by the gem structure and the course of action of iotas or particles inside the material. The presence of attractive anisotropy can influence the solidness of attractive spaces and the generally speaking attractive way of behaving of a material.

With regards to electron conduct, attractive anisotropy emerges from the association between the attractive snapshots of electrons and the precious stone grid. The direction of the precious stone grid can impact the favored course of arrangement for electron turns, prompting anisotropic attractive properties.

Spintronics and Magnetoelectronics:

The comprehension of the job of electrons in attraction has made ready for the advancement of spintronics (turn transport hardware) and magnetoelectronics. Not at all like conventional hardware, which depend on the charge of electrons, spintronics outfits the characteristic twist of electrons for data handling and stockpiling. This field of exploration can possibly upset electronic gadgets by offering benefits, for example, lower power utilization and higher information stockpiling thickness.

One of the vital parts in spintronics is the twist valve, a gadget that takes advantage of the twist subordinate conductivity of materials. In a twist valve, the overall direction of attractive minutes in two ferromagnetic layers decides the progression of twist energized electrons. This takes into account the control of electron turns and the control of electrical flows in view of their twist direction.

Magnetoresistance, the adjustment of electrical obstruction of a material because of an applied attractive field, is a peculiarity integral to spintronics. Goliath magnetoresistance (GMR) and burrow magnetoresistance (TMR) are two kinds of magnetoresistance that have been widely examined and applied in spintronic gadgets. These impacts depend on the arrangement or misalignment of attractive minutes in multi-facet structures.

Uses of Electron Attraction:

The comprehension of the job of electrons in attraction has prompted a great many mechanical applications. Attractive materials and gadgets assume basic parts in different fields, including data innovation, medication, and energy.

Attractive Capacity Gadgets:

Attractive capacity gadgets, for example, hard plate drives (HDDs) and attractive tapes, depend on the standards of electron attraction. The double data (0s and 1s) is encoded as the direction of attractive areas inside the capacity medium. The capacity to control and peruse these attractive spaces is fundamental for information capacity and recovery.

Attractive Reverberation Imaging (X-ray):

In the field of medication, X-ray machines use the standards of electron attractive reverberation to create point by point pictures of interior designs in the human body. The attractive properties of hydrogen cores in water particles are tested, giving important data about tissues and organs without the utilization of ionizing radiation.

Attractive Sensors:

Attractive sensors are generally utilized in different applications, including route frameworks, car innovation, and modern cycles. Lobby impact sensors, for instance, recognize changes in attractive fields and convert them into electrical signs. These sensors assume a pivotal part in place detecting and control frameworks.

Spintronic Gadgets:

Spintronic gadgets, including turn valves and attractive passage intersections, are at the front of spintronics research. These gadgets hold the potential for additional effective and smaller electronic parts, adding to the advancement of cutting edge PCs and memory gadgets.

Attractive Refrigeration:

The field of magnetocaloric materials investigates the utilization of attractive properties for cooling applications. Attractive refrigeration depends on the adjustment of temperature that happens when an attractive material is exposed to a changing attractive field. This harmless to the ecosystem innovation can possibly supplant customary gas pressure refrigeration frameworks.

Difficulties and Future Headings:

While how we might interpret the job of electrons in attraction has progressed fundamentally, there are still moves and secrets to be investigated. One area of continuous exploration is the journey for materials with improved attractive properties for applications in spintronics and attractive capacity. The improvement of new attractive materials with customized properties requires a profound comprehension of the hidden quantum mechanics.

One more test is the investigation of extraordinary attractive peculiarities, like topological covers and skyrmions. Topological separators are materials that show remarkable electronic states on their surfaces, and they hold guarantee for novel spintronic applications. Skyrmions, then again, are attractive quasiparticles that could be used for data capacity and handling in later advances.

Headways in trial procedures, like high-goal electron microscopy and high level spectroscopy, are empowering analysts to test attractive peculiarities at the nanoscale. The capacity to control and control individual attractive minutes opens up additional opportunities for the plan of nanoscale attractive gadgets with uncommon accuracy.

2.3 Ferromagnetism explained: the alignment of magnetic domains in iron

Ferromagnetism is an interesting peculiarity that has spellbound researchers and designers for a really long time. At the core of this attractive conduct lies the multifaceted arrangement of minute attractive spaces inside a material. Iron, a notable ferromagnetic material, fills in as a magnificent contextual analysis to investigate the standards of ferromagnetism and the arrangement of attractive spaces. In this investigation, we will dive into the central ideas of ferromagnetism, the job of electron turns, the arrangement of attractive spaces, and the mechanical ramifications of this attractive peculiarity.

Fundamental Standards of Ferromagnetism:

Ferromagnetism is a kind of attraction described by the unconstrained arrangement of attractive minutes in a material, prompting the making of naturally visible attractive spaces. The way to ferromagnetism lies in the quantum mechanical way of behaving of electrons, explicitly their characteristic property known as twist.

Electron turn is a key property that can be imagined as the revolution of an electron about its own pivot, creating an attractive second. The Pauli Rejection Guideline directs that no two electrons in a molecule can have a similar arrangement of quantum numbers, including turn. This guideline leads to the idea of electron turn states: "up" and "down."

In ferromagnetic materials, adjoining molecules or particles will generally have equal arrangement of electron turns, bringing about a net attractive second at the nuclear level. This arrangement of twists makes serious areas of strength for an attractive impact, making the material ferromagnetic. The trade connection, a quantum-mechanical impact emerging from the cross-over of electron wave capabilities, assumes a vital part in leaning toward the equal arrangement of twists in ferromagnetic materials.

Development of Attractive Spaces:

To comprehend ferromagnetism in materials like iron, investigating the idea of attractive domains is fundamental. Attractive spaces are locales inside a material where the attractive snapshots of iotas or particles are adjusted in a particular heading. Without an outside attractive field, these spaces might have irregular directions, bringing about a net plainly visible polarization of nothing.

At the point when an outside attractive field is applied to a ferromagnetic material, the attractive spaces will more often than not line up with the field. As the strength of the outside field builds, an ever increasing number of areas adjust, prompting a general expansion in the material's polarization. This arrangement cycle is a powerful balance, as nuclear power persistently contends with the outside field's impact, causing consistent vacillations in space direction.

The idea of attractive spaces was first proposed by Pierre-Ernest Weiss in 1907, and his model is known as the Weiss sub-atomic field hypothesis. As per this hypothesis, neighboring attractive minutes inside a material collaborate with one another, subsequent in a sub-atomic field that will in general adjust the attractive minutes parallelly. This arrangement makes attractive spaces, each acting like a minuscule magnet.

Ferromagnetic materials frequently display a trademark include known as immersion polarization. Immersion happens when an adequately solid outside attractive field adjusts every one of the attractive minutes in the material, and further expansions in the field strength don't prompt extra arrangement. The immersion charge is a critical boundary in describing the attractive way of behaving of ferromagnetic materials.

Hysteresis:

One of the particular ways of behaving related with ferromagnetic materials is hysteresis. Hysteresis is the slacking of the attractive enlistment (B) behind the attractive field strength (H) in a material exposed to a changing outside attractive field. This peculiarity is portrayed by a hysteresis circle in the B-H bend.

The B-H bend represents the connection between the attractive acceptance (B) and the attractive field strength (H) during the course of polarization and demagnetization. In the underlying phases of polarization, the attractive enlistment falls behind the expansion in attractive field strength because of the opposition of attractive spaces to adjust.

During the demagnetization cycle, the attractive acceptance falls behind the lessening in attractive field strength as the attractive areas oppose getting back to their unique irregular directions. The width of the hysteresis circle is a proportion of the energy misfortune in the material during the charge and demagnetization processes.

The coercivity of a ferromagnetic material is a key boundary connected with hysteresis. Coercivity addresses the attractive field strength expected to lessen the attractive acceptance to zero during the demagnetization cycle. Materials with high coercivity are more impervious to demagnetization and are frequently liked for long-lasting magnet applications.

Innovative Applications:

The comprehension of ferromagnetism and the arrangement of attractive spaces has prepared for various innovative applications. Here are a few prominent models:

Super durable Magnets:

Ferromagnetic materials, especially those with high immersion polarization and coercivity, are utilized to make extremely durable magnets. These magnets track down applications in different gadgets, including electric engines, generators, speakers, and attractive locks.

Attractive Recording:

The standards of ferromagnetism are principal to attractive recording advances, for example, hard plate drives (HDDs) and attractive tapes. In these gadgets, data is put away as polarized spaces on a ferromagnetic medium. The capacity to exactly control and control the arrangement of attractive spaces takes into account the encoding and recovery of information.

Transformers and Inductors:

Ferromagnetic materials are utilized in transformers and inductors to improve their attractive properties. The presence of a ferromagnetic center expands the inductance of the gadget, making it more proficient in the exchange of electrical energy.

Attractive Sensors:

Ferromagnetic materials are utilized in different attractive sensor advancements. Corridor impact sensors, for instance, distinguish changes in attractive fields and are utilized in applications, for example, position detecting, speed recognition, and current estimation.

Electric Engines and Generators:

The activity of electric engines and generators depends on the association between attractive fields. Ferromagnetic materials assume a pivotal part in upgrading the attractive properties of the center parts, working on the effectiveness and execution of these gadgets.

Difficulties and Advances in Ferromagnetic Exploration:

While ferromagnetism has been broadly examined and applied, there are progressing difficulties and areas of examination that keep on enamoring established researchers. Understanding and controlling the elements of attractive spaces at the nanoscale is a vital focal point of examination, driven by the journey for more modest and more productive attractive gadgets.

Nanoscale Attraction:

Propels in nanotechnology have opened up additional opportunities for controlling and controlling attractive areas at the nanoscale. Scientists are investigating methods to make nanoscale attractive designs with explicit properties, prompting progressions in spintronics and the improvement of novel attractive materials.

Topological Separators:

The investigation of topological separators has uncovered charming attractive properties. These materials display novel electronic states on their surfaces and are being explored for expected applications in spintronics. The disclosure of topological separators has extended the comprehension of quantum attractive peculiarities.

Skyrmions:

Skyrmions are attractive quasiparticles that definitely stand out. These steady, vortex-like designs can be controlled at the nanoscale and have possible applications in data capacity and handling. The investigation of skyrmionic frameworks addresses a boondocks in ferromagnetic examination.

Energy-Effective Attractive Materials:

The interest for energy-productive advances has prodded investigation into attractive materials with decreased energy misfortunes during charge and demagnetization processes. Creating materials with lower hysteresis and further developed attractive properties is a critical concentration for applications in power hardware and environmentally friendly power frameworks.

Chapter 3

The Quantum World Of Magnetism

Attraction, a peculiarity well established in the quantum world, has been a subject of interest and investigation for researchers from the beginning of time. At the quantum level, the way of behaving of rudimentary particles, especially electrons, assumes a critical part in deciding the attractive properties of materials. In this broad investigation, we will dig into the quantum mechanical rules that underlie attraction, the quantum idea of electron turns, attractive minutes, and the different attractive peculiarities saw in materials. From understanding the twist fermion communication to investigating quantum turn fluids, this excursion into the quantum universe of attraction will uncover the complicated associations between quantum mechanics and attractive peculiarities.

1. **Quantum Mechanics and Electron Twist:**

   At the core of attraction in the quantum world lies the natural property of electrons known as twist. In the structure of quantum mechanics, the way of behaving of electrons is depicted not just by their orbital movement around a core yet in addition by their twist, a natural rakish force. Not at all like old style turning objects, electron turn is a quantum property, and its quantization prompts exceptional attractive qualities.

   The Pauli Prohibition Standard, a basic rule of quantum mechanics, directs that no two electrons in a molecule can have a similar arrangement of quantum numbers, including turn. This rule presents the idea of electron turn states: "up" and "down." The twist of an electron creates an attractive second, similar to a small bar magnet, with the heading existing apart from everything else lined up with the twist.

   The Harsh Gerlach analyze, directed in 1922, gave exploratory proof to the quantization of electron turn. The investigation uncovered that the attractive snapshots of electrons are quantized along a particular course, adjusting either with or against an applied attractive field. This quantization of twist has significant ramifications for the attractive properties of materials at the quantum level.

2. **Quantum Twist Circle Coupling:**

Notwithstanding electron turn, another quantum mechanical impact that essentially impacts attraction is turn circle coupling. Turn circle coupling emerges from the collaboration between the attractive second connected with an electron's twirl and its orbital movement around the core. This connection prompts a coupling between the electron's twist and its orbital rakish energy. The relativistic impacts engaged with turn circle coupling bring about the lifting of declines in the energy levels of electrons with various quantum numbers. This lifting of declines can impact the attractive properties of materials, prompting peculiarities like attractive anisotropy. Attractive anisotropy alludes to the reliance of a material's attractive properties on the course of the applied attractive field, and it assumes a vital part in deciding the favored direction of attractive minutes inside a material.

Understanding twist circle coupling is especially significant in materials with weighty components, where relativistic impacts become more articulated. The interchange between electron turn and orbital movement brings about complex attractive way of behaving that can't be completely made sense of by considering turn alone.

3. **Quantum Twist Elements:**

In the quantum world, the elements of electron turns are administered by quantum turn Hamiltonians, numerical portrayals of the energy related with the collaborations between turns in a material. The Heisenberg model, a fundamental quantum turn Hamiltonian, depicts the trade communication between adjoining turns, affecting the arrangement of attractive minutes.

The trade collaboration is a quantum-mechanical impact emerging from the cross-over of electron wave capabilities. At the point when two electrons are near one another, their wave capabilities cross-over, prompting a trade of their spatial directions. The trade collaboration can be either ferromagnetic, inclining toward equal arrangement of twists, or antiferromagnetic, leaning toward antiparallel arrangement.

The Heisenberg model structures the reason for grasping different attractive peculiarities, including ferromagnetism, antiferromagnetism, and ferrimagnetism. Quantum turn elements likewise presents ideas like quantum variances, where the direction of twists vacillates because of the vulnerability rule of quantum mechanics. Quantum variances become more articulated at low temperatures, prompting peculiarities like quantum turn fluids.

4. **Ferromagnetism and Quantum Trade Collaboration:**

Ferromagnetism, portrayed by the unconstrained arrangement of attractive minutes in a material, is a sign of the quantum trade collaboration. In ferromagnetic materials, adjoining turns will generally adjust lined up with one another, subsequent in a solid by and large attractive impact. The trade collaboration, especially the trade energy term in the Heisenberg model, assumes a pivotal part in settling the ferromagnetic request.

The beginning of ferromagnetism lies in the arrangement of twists, driven by the trade collaboration leaning toward equal direction. The higher the trade energy, the more steady the ferromagnetic request becomes. Materials with an enormous trade energy, like iron and cobalt, display solid ferromagnetic way of behaving.

Quantum precisely, the trade communication is an indication of the Pauli Rejection Rule, which disallows two electrons from involving a similar quantum state. At the point when electrons have equal twists, the trade cooperation is positive, prompting a lower energy state. This quantum impact brings about the adjustment of equal twist arrangement in ferromagnetic materials.

5. **Antiferromagnetism and Quantum Dissatisfaction:**

Antiferromagnetism, rather than ferromagnetism, includes the arrangement of twists in inverse headings. In antiferromagnetic materials, adjoining turns will quite often adjust antiparallel to one another. The quantum trade cooperation in antiferromagnetic materials prompts the abrogation of the generally attractive second at the perceptible level.

Quantum dissatisfaction is a vital idea in grasping antiferromagnetism. Dissatisfaction emerges when it is absurd to expect to fulfill every one of the collaborations between adjoining turns all the while, prompting contending propensities in the arrangement of twists. This dissatisfaction can bring about remarkable attractive ground states, like twist fluids, where the twists stay disarranged even at extremely low temperatures.

The disappointment in antiferromagnetic frameworks can be mathematically actuated, for example, in three-sided or Kagome grids, where the cross section calculation forestalls the synchronous fulfillment of all trade connections. The investigation of disappointed magnets has turned into a boondocks in quantum attraction, investigating outlandish attractive stages and quantum turn fluids.

6. **Quantum Twist Fluids:**

Quantum turn fluids address a novel and captivating period of issue in which twists stay confused down to the most minimal temperatures. In contrast to conventional attractive stages, where twists adjust in arranged designs, turn fluids are described by powerful and fluctuating twists. Quantum variances, emerging from the Heisenberg vulnerability rule, forestall the foundation of long-range attractive request in turn fluids.

The idea of quantum turn fluids was at first proposed by Nobel laureate Philip W. Anderson in 1973. In a twist fluid, the ground state is exceptionally degenerate, considering a large number of twist designs. This decadence is a result of dissatisfaction, and the ground state stays dynamic, keeping away from the foundation of a special attractive request.

Quantum turn fluids have been proposed and concentrated on in different materials, including specific sorts of natural mixtures and certain mathematically baffled magnets. The journey for noticing and understanding quantum turn fluids is driven by their likely importance in the more extensive field of quantum dense matter material science and their suggestions for quantum figuring.

7. **Quantum Stage Changes:**

Quantum stage changes are peculiarities that happen at outright zero temperature and are driven by quantum vacillations. Dissimilar to traditional stage changes that happen at limited temperatures, quantum stage changes are related with the tuning of a non-warm boundary, like attractive field strength or strain.

The Landau hypothesis of stage advances, created for traditional frameworks, is deficient for depicting quantum stage changes. Quantum stage advances include the opposition between various ground states, frequently determined by quantum changes and the fragile harmony between contending associations.

The investigation of quantum deliberately work advances in attractive frameworks has uncovered captivating peculiarities, for example, the development of new attractive stages and basic way of behaving related with the quantum basic point. Quantum criticality, happening at the progress point, is portrayed by the shortfall of an energy scale and the commonness of all inclusive scaling regulations.

8. **Topological Magnets and Quantum Inconsistencies:**

Topological periods of issue have turned into a front in the investigation of quantum attraction. Topological magnons, semi particles related with aggregate twist excitations, have been proposed and seen in specific attractive materials. These topological magnons can display safeguarded edge states, like the edge states in topological encasings, making them powerful against neighborhood irritations.

Quantum oddities, infringement of old style balances at the quantum level, have been recognized in topological magnets. The chiral oddity, for instance, appears as the non-protection of the quantity of chiral charge transporters within the sight of attractive fields. The exchange among geography and quantum oddities opens up new roads for understanding and controlling attractive properties at the quantum level.

9. **Quantum Processing and Spintronics:**

The marriage of quantum mechanics and attraction has huge ramifications for the fields of quantum processing and spintronics. Quantum figuring, utilizing the standards of quantum superposition and entrapment, holds the potential for taking care of specific issues dramatically quicker than old style PCs.

Quantum bits or qubits, the fundamental units of quantum data, can be carried out utilizing the twist conditions of electrons. Attractive materials with long intelligibility times and clear cut turn states are critical for the improvement of stable qubits. Also, the idea of quantum entrapment can be bridled for quantum correspondence and quantum cryptography.

Spintronics, a field that takes advantage of the twist of electrons for data handling and stockpiling, has arisen as a promising innovation. Gadgets, for example, turn valves and attractive passage intersections use the quantum properties of electron turns for controlling and recognizing data. Quantum lucidness in spintronic gadgets is fundamental for accomplishing elite execution and low-energy utilization.

10. **Trial Methods in Quantum Attraction:**

Propels in exploratory methods have been instrumental in investigating the quantum universe of attraction. High-goal neutron dissipating and X-beam diffraction permit scientists to test the attractive design of materials with choice accuracy. Inelastic neutron dispersing gives bits of knowledge into the powerful properties of twist excitations, assisting specialists with grasping the quantum elements of attractive frameworks.

Atomic attractive reverberation (NMR) and electron paramagnetic reverberation (EPR) spectroscopy are integral assets for researching the neighborhood attractive properties of materials. These procedures permit specialists to concentrate on the way of behaving of individual cores or unpaired electrons, giving important data about the electronic construction and attractive associations.

Filtering burrowing microscopy (STM) and nuclear power microscopy (AFM) empower scientists to picture and control individual iotas and attractive designs on surfaces. These methods play had a pivotal impact in the investigation of quantum turn frameworks at the nanoscale.

3.1 Quantum mechanics and its impact on our understanding of magnetism

Quantum mechanics, the progressive hypothesis created in the mid twentieth 100 years, has significantly impacted how we might interpret the crucial powers administering the way of behaving of issue at the littlest scales. Among the different peculiarities enlightened by quantum mechanics, attraction stands apart as a complicated and captivating area of study. In this investigation, we will dig into the vital standards of quantum mechanics that have formed how we might interpret attraction, looking at the quantum idea of electron turns, the job of trade associations, and the rise of novel attractive peculiarities. From the origin of quantum mechanics to late headways in the field, the effect on our perception of attraction has been extraordinary.

1. **Electron Twist and Inherent Attractive Minutes:**

   At the core of quantum mechanics lies the idea of electron turn, a quantum property that recognizes it from old style turning objects. In the mid 1920s, the advancement of quantum mechanics presented the possibility that electrons have a natural rakish force, or twist. Dissimilar to old style rakish energy, electron turn is quantized, and it brings about an inherent attractive second connected with the turning electron.

   The quantization of twist prompts the polarity of twist states: "up" and "down." This intrinsic property of electrons adds to the development of attractive minutes at the quantum level. The Pauli Rejection Standard directs that no two electrons inside a particle can have a similar arrangement of quantum numbers, including turn. This rule, joined with the quantization of twist, lays out the establishment for the exceptional attractive conduct saw in materials.

   The Harsh Gerlach try, led in 1922, gave direct exploratory proof of quantized electron turn. The examination included passing a light emission molecules through an inhomogeneous attractive field, making the shaft split into discrete spots on a finder screen. The noticed quantization of electron turn affirmed the presence of natural attractive minutes related with electrons, making way for the investigation of attraction in the quantum domain.

2. **Quantum Superposition and Trap:**

   Two other key standards of quantum mechanics, superposition, and trap, assume a significant part in molding how we might interpret attraction. Superposition permits quantum particles to exist in numerous states at the same time, testing traditional instinct. With regards to attraction, superposition turns out to be especially applicable while considering the direction of electron turns.

   In a quantum superposition of twist expresses, an electron twist can exist in a mix of "up" and "down" states until an estimation is made, falling the superposition into one of the potential states. This quantum highlight is taken advantage of in quantum registering, where qubits can exist in superpositions of 0 and 1, giving a parallelism not feasible in old style processing.

   Entrapment, one more one of a kind quantum peculiarity, depicts major areas of strength for the between particles, in any event, when isolated by enormous distances. With regards to attraction, snare can happen between the twists of particles, making a corresponded express that perseveres no matter what the spatial partition between particles.

Quantum entrapment has been tentatively exhibited in frameworks of ensnared electron turns. The ensnarement of twists in a material can prompt aggregate way of behaving, impacting the attractive properties of the whole framework. The investigation of snare in attractive frameworks has opened up new roads for grasping quantum relationships and their effect on new attractive peculiarities.

3. **Quantum Mechanics of Attractive Materials:**

The way of behaving of attractive materials is on a very basic level represented by the quantum mechanical properties of electrons inside the material. Quantum mechanics presents the idea of wave capabilities, which depict the likelihood dissemination of tracking down an electron in a specific state. The spatial dissemination of electron wave capabilities assumes a pivotal part in grasping the plan of electrons in attractive materials.

In attractive materials, the attractive minutes related with individual electron turns associate with one another, prompting the development of aggregate attractive way of behaving. The trade connection, a quantum mechanical impact emerging from the cross-over of electron wave capabilities, is a critical figure deciding the arrangement of twists in attractive materials.

The Heisenberg model, a quantum turn Hamiltonian, is frequently used to depict the trade collaboration between adjoining turns. In this model, the trade communication term directs whether twists like to adjust equal (ferromagnetic) or antiparallel (antiferromagnetic). Quantum mechanics gives the hypothetical structure to figuring out the energy commitments and soundness of various attractive plans inside a material.

4. **Quantum Twist Elements and Twist Waves:**

Quantum mechanics likewise gives bits of knowledge into the powerful way of behaving of attractive materials, especially with regards to turn elements. The quantum turn Hamiltonian incorporates terms that portray the associations among turns and their dynamic advancement over the long haul.

Turn waves, otherwise called magnons, are aggregate excitations of twists in an attractive material. These excitations can be depicted quantum precisely, regarding the twists as quantized substances. The quantization of twist takes into consideration the creation and obliteration of magnons, giving a quantum viewpoint on the spread of twist waves through a material.

Quantum turn elements, represented by the Heisenberg model, uncovers the energy scattering connection of twist waves, showing how their energy relies upon their wave vector. Trial procedures, for example, inelastic neutron dispersing, have been instrumental in straightforwardly noticing and examining magnons, affirming the quantum idea of twist excitations in attractive materials.

5. **Trade Connection and Attractive Stages:**

The trade connection, a quantum mechanical impact emerging from the Pauli Prohibition Standard, essentially impacts the attractive properties of materials. The idea of the trade connection decides the sort of attractive request that arises in a material.

Ferromagnetism: In ferromagnetic materials, the trade cooperation favors equal arrangement of twists, prompting the unconstrained charge of the material. At the quantum level, the trade connection brings down the energy of the framework when twists adjust in a similar course. This arrangement perseveres even without any an outside attractive field, bringing about the development of ferromagnetic spaces.

Antiferromagnetism: In antiferromagnetic materials, the trade connection favors antiparallel arrangement of twists, bringing about an undoing of the in general attractive second. The quantum mechanics of antiferromagnetism includes contending communications between turns, prompting the development of attractive areas with antiparallel twist directions.

Ferrimagnetism: Ferrimagnetic materials show a more perplexing attractive construction, where two sublattices with various attractive minutes coincide. The quantum idea of ferrimagnetism includes an unevenness in the extents of the attractive minutes, prompting a net attractive second for the whole material.

The comprehension of these attractive stages at the quantum level is essential for anticipating and controlling the attractive properties of materials for different applications.

6. **Quantum Oddities and Topological Angles:**

Quantum oddities, infringement of traditional balances at the quantum level, have been distinguished with regards to attractive materials. The chiral peculiarity, for example, is a quantum irregularity related with the non-protection of the quantity of chiral charge transporters within the sight of attractive fields. Quantum irregularities give a novel viewpoint on the exchange among geography and attraction.

Topological parts of attraction have turned into a front in quantum research. Topological separators, which display exceptional electronic states on their surfaces, have analogs in the field of topological magnons. The investigation of topological magnons includes the investigation of safeguarded edge states in attractive materials, offering possible applications in spintronics and quantum data handling.

7. **Quantum Twist Fluids:**

Quantum turn fluids address an outlandish and slippery condition of issue where twists stay scattered down to the least temperatures. In a quantum turn fluid, the ground state is profoundly ruffian, and twists are in a dynamic and fluctuating state, sidestepping the foundation of long-range attractive request. The idea of quantum turn fluids challenges customary thoughts of attractive request and has been a subject of serious hypothetical and trial examination. Quantum changes, emerging from the Heisenberg vulnerability rule, keep the twists from sinking into a static attractive setup. The investigation of quantum turn fluids holds guarantee for uncovering new periods of issue and figuring out the crucial furthest reaches of attractive request.

8. **Quantum Stage Changes:**

Quantum stage changes, happening at outright zero temperature and driven by quantum variances, address a captivating part of quantum attraction. Not at all like old style stage changes, quantum stage advances are related with the tuning of non-warm boundaries, like attractive field strength or tension.

The Landau hypothesis of stage changes, produced for old style frameworks, is insufficient for portraying quantum stage advances. Quantum criticality, related with a quantum stage change, is described by the shortfall of an energy scale and the pervasiveness of widespread scaling regulations. The investigation of quantum stage advances gives bits of knowledge into the opposition between various attractive ground states and the rise of novel quantum conditions of issue.

9. **Innovative Ramifications and Quantum Attraction:**

The effect of quantum mechanics on how we might interpret attraction reaches out past hypothetical investigation to down to earth applications and innovative headways. Quantum registering, which takes advantage of the standards of quantum superposition and entrapment, holds the possibility to alter the field of computational attraction.

Quantum bits or qubits, executed utilizing the quantum conditions of electron turns, can play out specific estimations dramatically quicker than traditional PCs. Quantum calculations for reproducing attractive materials and investigating attractive stage advances are effectively being created, opening additional opportunities for understanding and planning materials with custom-made attractive properties.

In the domain of spintronics, quantum mechanics assumes an essential part in the improvement of gadgets that use the twist of electrons for data handling and stockpiling. Turn valves, attractive passage intersections, and other spintronic gadgets influence the quantum properties of electron twists to control and identify data with high effectiveness and low energy utilization.

3.2 Spin and magnetic moments: the microscopic view of magnetic behavior

Attraction is a peculiarity that has enthralled human interest for a really long time. From the early utilization of lodestones in old China to the improvement of present day attractive advances, the comprehension of attraction has developed fundamentally. At the minute level, the way of behaving of attractive materials is unpredictably attached to the idea of twist and attractive minutes. In this investigation, we dive into the minute domain to unwind the intricacies of twist and attractive minutes, acquiring experiences into the entrancing universe of attractive way of behaving.

Twist and Quantum Mechanics:

The minute way of behaving of attractive materials tracks down its underlying foundations in the quantum mechanical properties of electrons. Quantum mechanics changed how we might interpret particles at the nuclear and subatomic levels. One vital part of quantum mechanics is the idea of electron turn. Turn is an inherent property of electrons that invests them with a natural precise energy, similar as a minuscule turning top.

The twist of an electron is portrayed by a quantum number, and it can take one of two potential qualities: up or down. This paired nature of electron turn has significant ramifications for attractive way of behaving. At the point when electrons possess nuclear orbitals, their twists will quite often adjust in unambiguous ways, leading to attractive minutes.

Attractive Minutes:

An attractive second is a vector amount that portrays the strength and direction of the attractive properties of an item. With regards to molecules and materials, attractive minutes emerge because of the attractive idea of electrons. The most major attractive second in materials comes from the natural attractive dipole second connected with the electron turn.

The direction of the attractive snapshot of an electron can be envisioned as a bolt pointing along its pivot. At the point when numerous attractive minutes are available in a material, their aggregate way of behaving decides the material's attractive properties. Understanding the beginning and game plan of these attractive minutes is significant for disentangling the tiny components fundamental attractive way of behaving.

Trade Connection:

In the tiny universe of attractive materials, communications between adjoining attractive minutes assume a critical part. The trade collaboration is a quantum mechanical peculiarity that oversees the arrangement of twists in a material. It emerges because of the covering of electron wave capabilities in contiguous iotas or particles.

There are two primary kinds of trade communications: ferromagnetic and antiferromagnetic. In ferromagnetic materials, adjoining attractive minutes will generally adjust lined up with one another, subsequent in a plainly visible polarization of the material. This arrangement is answerable for the recognizable properties of magnets, for example, drawing in or repulsing different magnets.

Then again, in antiferromagnetic materials, adjoining attractive minutes adjust in inverse headings, prompting an undoing of their plainly visible attractive impacts. This outcomes in remarkable attractive ways of behaving, and antiferromagnetic materials are many times utilized in mechanical applications, like attractive stockpiling gadgets.

Quantum Mechanical Models:

To depict the attractive properties of materials at the tiny level, quantum mechanical models are utilized. One such model is the Heisenberg model, which thinks about the connection between attractive minutes on adjoining destinations inside a gem cross section. The Heisenberg model gives a structure to understanding what the trade communication means for the attractive way of behaving of a material.

The model presents a trade boundary that evaluates the strength of the connection between turns. Contingent upon the sign and size of this boundary, the material can display ferromagnetic, antiferromagnetic, or much more intricate attractive ways of behaving. The Heisenberg model has been instrumental in explaining the attractive properties of different materials and foreseeing their conduct under various circumstances.

Job of Gem Construction:

The plan of particles in a material's gem cross section likewise assumes a pivotal part in deciding its attractive properties. The gem structure impacts the spatial game plan of attractive minutes, influencing how they interface with one another. In certain materials, the precious stone construction might lean toward specific attractive arrangements, prompting the development of explicit attractive stages.

For instance, in materials with a cubic precious stone construction, there might be a propensity for the attractive minutes to adjust along unambiguous tomahawks, leading to attractive anisotropy. This anisotropy can impact the material's reaction to outside attractive fields and is a vital figure the plan of attractive materials for different applications.

Temperature and Warm Changes:

The tiny way of behaving of attractive minutes isn't static yet is dependent upon warm vacillations. As temperature expands, the nuclear power disturbs the arranged arrangement of twists, prompting a decrease in plainly visible polarization. The basic temperature at which a material goes through a progress from an attractive to a non-attractive state is known as the Curie temperature.

Over the Curie temperature, a ferromagnetic material loses its unconstrained polarization, becoming paramagnetic. In the paramagnetic express, the attractive minutes are haphazardly situated, and the material doesn't show a net plainly visible polarization. Figuring out the interchange between temperature, warm changes, and attractive way of behaving is fundamental for the plan and streamlining of attractive materials for explicit applications.

Quantum Burrowing of Charge:

In specific attractive materials, the way of behaving of attractive minutes goes past traditional depictions and enters the domain of quantum mechanics. Quantum burrowing of charge is a peculiarity where the attractive snapshot of a material can progress between various energy states through quantum burrowing.

This peculiarity is especially important in the investigation of nanoscale attractive particles, where the impacts of quantum mechanics become articulated. Quantum burrowing of polarization has suggestions for the advancement of super delicate attractive sensors and quantum data handling gadgets, featuring the rich transaction between quantum mechanics and attraction at the minuscule level.

Spintronics: Saddling Twist for Innovation:

The comprehension of twist and attractive minutes has prompted the development of spintronics, a field that investigates the control and use of electron turn for mechanical applications. Not at all like customary gadgets, which depend on the charge of electrons, spintronics outfits the twist of electrons to encode and handle data.

One critical component in spintronics is the improvement of spintronic gadgets, like twist valves and attractive passage intersections. These gadgets influence the capacity to control and identify the twist direction of electrons, empowering the production of additional effective and adaptable electronic parts. Spintronics can possibly upset registering and data capacity, offering new roads for energy-proficient and superior execution gadgets.

3.3Quantum theories and their applications in explaining iron's magnetism

Attraction, a peculiarity profoundly implanted in our ordinary encounters, tracks down its foundations in the quantum mechanical properties of issue. The comprehension of attractive way of behaving has gone through a groundbreaking excursion, pushed by the coming of quantum speculations. Iron, a typical component with unprecedented attractive properties, fills in as a captivating subject to investigate the significant ramifications of quantum mechanics in clarifying the tiny beginnings of attraction. In this extensive examination, we dig into the multifaceted quantum hypotheses overseeing iron's attraction and investigate their boundless applications in both hypothetical exploration and mechanical headways.

1. **Quantum Mechanics and Electron Twist:**

   1.1 The Quantum Mechanical Structure:

   The underpinning of quantum speculations in attraction lies in the standards of quantum mechanics, a system that upset comprehension we might interpret the way of behaving of particles at the nuclear and subatomic scales. Quantum mechanics presented a takeoff from traditional material science, supplanting deterministic directions with probabilistic wave capabilities.

   1.2 Electron Twist and Inherent Rakish Force:

At the core of quantum attraction is the idea of electron turn. Electrons, as rudimentary particles, have an inherent property known as twist, a quantum mechanical rakish force. Dissimilar to old style turning objects, electron turn is intrinsically quantum in nature, complying to the principles of quantum measurements.

The quantization of electron turn prompts a twofold property - electrons can have turn upsides of by the same token "up" or "down." This double nature in a general sense impacts the attractive properties of materials, as the aggregate way of behaving of electron turns leads to naturally visible polarization.

1.3 Job of Electron Twist in Iron's Attraction:

In iron, a change metal, the quantum mechanical nature of electron turn assumes an essential part in the rise of ferromagnetism. Each iron particle contributes numerous electrons, and their twists will more often than not adjust, bringing about a net attractive second for the whole material. This arrangement of electron turns is essential for the appearance of ferromagnetic properties in iron.

Understanding the quantum mechanics of electron turn gives a hypothetical system to foreseeing and controlling attractive properties at the nuclear level, laying the basis for additional investigation.

2. Trade Communication and Ferromagnetism:

2.1 The Beginning of Trade Connection:

The trade connection, a quantum mechanical peculiarity, is fundamental in making sense of the arrangement of electron turns in attractive materials. It emerges from the Pauli avoidance guideline, expressing that no two electrons can possess a similar quantum state at the same time. Thus, electrons in adjoining molecules or particles will more often than not try not to be in a similar twist state.

2.2 Ferromagnetism in Iron:

In ferromagnetic materials like iron, the trade communication favors equal arrangement of adjoining electron turns. At the point when electrons in nearby molecules have equal twists, the trade association turns out to be vivaciously good. This arrangement stretches out over plainly visible scales, prompting the development of attractive spaces.

The helpful arrangement of twists in ferromagnetic materials leads to an unconstrained naturally visible polarization, a critical trait of ferromagnetism. This arrangement is supported by the trade communication, a quantum mechanical rule that oversees the way of behaving of electron turns for a minuscule scope.

2.3 Temperature Reliance and Curie Temperature:

The quantum mechanical nature of the trade cooperation is fundamental for figuring out the temperature reliance of ferromagnetic materials. As temperature increments, nuclear power disturbs the arranged arrangement of twists, causing a decrease in perceptible polarization.

The basic temperature at which a ferromagnetic material goes through a progress from an attractively requested to a confused state is known as the Curie temperature. Quantum mechanics gives experiences into what warm vacillations mean for the solidness of attractive request in materials, prompting the deficiency of ferromagnetism over the Curie temperature.

3. **Precious stone Construction and Attractive Anisotropy:**

3.1 Impact of Gem Construction:

The glasslike game plan of particles in a material significantly impacts its attractive properties. On account of iron, the body-focused cubic (BCC) gem structure assumes an essential part in deciding the direction of attractive minutes.

3.2 Attractive Anisotropy:

Attractive anisotropy alludes to the directional reliance of a material's attractive properties. In iron, the BCC precious stone construction adds to the improvement of attractive anisotropy, implying that the material displays particular charge along certain crystallographic tomahawks.

Understanding attractive anisotropy is fundamental for fitting attractive materials to explicit applications. By controlling the gem structure, analysts can design materials with improved or smothered attractive anisotropy, affecting their reaction to outside attractive fields.

4. **Quantum Burrowing and Warm Variances:**

4.1 Quantum Burrowing of Charge:

At limited temperatures, warm variances assume a huge part in the way of behaving of attractive materials. Quantum burrowing of polarization, a quantum mechanical peculiarity, becomes pertinent in understanding how attractive minutes change between various energy states.

4.2 Nanoscale Peculiarities:

Quantum burrowing of charge is especially articulated in nanoscale attractive particles. As the size of attractive particles diminishes, the impacts of quantum mechanics become more conspicuous. Understanding quantum burrowing is urgent for anticipating and controlling the attractive way of behaving of nanoscale materials, with suggestions for both basic examination and innovative applications.

4.3 Ramifications for Attractive Capacity:

The transaction between warm variances and quantum burrowing has suggestions for attractive capacity innovation. In the plan of attractive stockpiling gadgets, for example, hard plate drives, the soundness of attractive states and the unwavering quality of data stockpiling are impacted by the sensitive harmony between warm impacts and quantum mechanical peculiarities.

5. **Spintronics and Mechanical Applications:**

5.1 Rise of Spintronics:

The marriage of quantum mechanics and attraction has brought about the field of spintronics, where the characteristic twist of electrons is saddled for data handling and stockpiling. Iron, with its surely known attractive properties, assumes a critical part in progressing spintronics examination and applications.

5.2 Spintronic Gadgets:

Spintronics presents a change in outlook from conventional hardware, depending on the control of electron charge, to a structure where electron turn is the essential transporter of data. Spintronic gadgets, for example, turn valves and attractive passage intersections, influence the control and identification of electron turn, offering benefits regarding energy proficiency and handling speed.

5.3 Quantum Spot Spintronics:

Quantum spots, nanoscale semiconductor particles, bring quantum mechanical impacts into spintronics. Quantum spot spintronics investigates the associations between bound electrons and their twists, opening up additional opportunities for quantum data handling and correspondence.

6. **Quantum Monte Carlo Recreations and Iron's Attractive Properties:**

6.1 Computational Apparatuses for Quantum Recreations:

Quantum Monte Carlo (QMC) recreations address a strong computational instrument for investigating the quantum mechanical parts of iron's attraction. These reenactments give a minute perspective on the electronic design of materials, taking into account the impacts of electron collaborations and other quantum peculiarities.

6.2 Foreseeing Attractive Properties:

QMC reproductions have been instrumental in foreseeing and figuring out different attractive properties of iron. From attractive stage advances to attractive anisotropy, these recreations offer experiences into the perplexing quantum communications overseeing iron's way of behaving.

6.3 Towards Materials Plan:

The use of QMC reproductions reaches out past hypothetical examinations, directing the plan of new materials with custom-made attractive properties. By incorporating quantum mechanical standards into computational models, specialists can investigate novel materials that show wanted attractive ways of behaving, adding to the progression of materials science.

Chapter 4

Magnetic Materials Beyond Iron

Attraction is an intriguing and principal part of physical science, with iron being quite possibly of the most notable attractive material. Nonetheless, the universe of attractive materials stretches out a long ways past iron, enveloping a different scope of substances with novel attractive properties. In this investigation, we dig into the domain of attractive materials past iron, talking about different kinds of magnets, their applications, and the state of the art research forming how we might interpret attraction.

Kinds of Attractive Materials

Ferromagnetic Materials

While iron is an exemplary model, different components and mixtures show ferromagnetism, where individual attractive minutes adjust in a similar course, bringing about areas of strength for an attractive second. Cobalt and nickel are outstanding instances of ferromagnetic components. Combinations like alnico (aluminum, nickel, and cobalt) and ferrites (fired compounds) are additionally usually utilized ferromagnetic materials.

Antiferromagnetic Materials

In antiferromagnetic materials, neighboring attractive minutes adjust in inverse headings, making their attractive impacts counterbalance one another. Manganese oxide is a notable antiferromagnetic material. Antiferromagnets are vital in attractive capacity applications, where data is put away in the general directions of adjoining attractive minutes.

Ferrimagnetic Materials

Ferrimagnetic materials consolidate parts of both ferromagnetism and antiferromagnetism. In these materials, two sublattices with various attractive minutes are available, and the net attractive second emerges from the distinction between the two. Blended metal oxides like magnetite (Fe_3O_4) display ferrimagnetic conduct.

Paramagnetic Materials

Not at all like ferromagnetic materials, paramagnetic materials have no inherent attractive second. Be that as it may, when presented to an outer attractive field, they foster a powerless initiated attractive second. Normal paramagnetic materials incorporate aluminum, platinum, and certain metal particles. Paramagnets are significant in fields like clinical imaging, where contrast specialists use their attractive properties.

Diamagnetic Materials

Diamagnetic materials, conversely, show a frail negative reaction to an outside attractive field. Most materials show some level of diamagnetism, however it is regularly extremely powerless. Superconductors, when in their superconducting state, areas of strength for show properties, removing attractive fields.

Utilizations of Attractive Materials Past Iron

Gadgets and Information Stockpiling

Attractive materials assume a crucial part in the hardware business, especially in information capacity. While hard drives generally utilized iron-based composites, current innovations consolidate various attractive materials. Opposite attractive recording (PMR) and the arising heat-helped attractive recording (HAMR) innovations influence progressed attractive materials to increment information capacity thickness and productivity.

Clinical Imaging

In the domain of medication, attractive reverberation imaging (X-ray) depends on the attractive properties of specific materials, frequently paramagnetic differentiation specialists, to produce nitty gritty pictures of interior body structures. The capacity to control and control attractive properties adds to the improvement of more productive and designated imaging methods.

Power Age and Transformation

Super durable magnets, frequently produced using neodymium-iron-boron (Nd-FeB) or samarium-cobalt (SmCo), are essential parts in electric generators and engines. These magnets, displaying solid attractive properties, empower the effective transformation of electrical energy into mechanical energy as well as the other way around. Propels in magnet innovation add to the improvement of additional proficient and minimized electric machines.

Attractive Sensors and Actuators

Attractive sensors, for example, Lobby impact sensors, depend on the reaction of materials to attractive fields. These sensors track down applications in different businesses, including auto (for position detecting), purchaser gadgets (as vicinity sensors), and modern robotization. Attractive actuators, driven by the collaboration of attractive fields, are vital to gadgets like solenoids and attractive valves.

Past Iron: Arising Attractive Materials

Uncommon Earth Magnets

While not totally past iron, uncommon earth magnets, especially those consolidating neodymium and samarium, address a critical headway in attractive materials. These magnets display outstanding strength, making them imperative in applications where minimization and high attractive execution are basic, like in electric vehicles and wind turbines.

Spintronics and Attractive Passage Intersections

Spintronics, short for turn transport gadgets, investigates the inherent twist of electrons for data capacity and handling. Attractive passage intersections (MTJs), made out of flimsy layers of attractive and non-attractive materials, are basic to spintronic gadgets. They are utilized in attractive irregular access memory (MRAM) and add to the improvement of more energy-effective gadgets.

Topological Protectors

Topological protectors are a class of materials that show special electronic properties on their surfaces. A few topological separators show attractive impacts, and scientists are investigating their true capacity for spintronic applications. The surface conditions of these materials could be bridled for dissemination free electronic vehicle, adding to the advancement of more hearty and energy-proficient hardware.

Attractive Fluids

Attractive fluids, otherwise called ferrofluids, comprise of nanoparticles suspended in a transporter liquid. These fluids answer attractive fields, showing interesting ways of behaving like shape-moving and controllable liquid elements. Ferrofluids have applications in advances going from mechanical seals and damping frameworks to biomedical applications like designated drug conveyance.

Difficulties and Future Headings

Ecological Effect of Intriguing Earth Magnets

While interesting earth magnets offer excellent attractive properties, their creation includes extricating and handling components with huge ecological effect. Specialists are investigating elective materials or reusing techniques to moderate the ecological outcomes of uncommon earth magnet creation.

Energy-Productive Spintronics

Progressing spintronics requires defeating difficulties connected with energy productivity and control of twist states. Creating materials with custom-made attractive properties and proficient twist control methods is significant for understanding the maximum capacity of spintronic gadgets.

Attractive Materials for Quantum Registering

The field of quantum registering depends on the control of quantum bits (qubits), and attractive materials assume a part in making steady and controllable qubits. Investigating the collaboration between attractive properties and quantum states is a continuous area of examination with the possibility to reform registering abilities.

Biomedical Utilizations of Attractive Materials

In medication, specialists are examining the utilization of attractive materials for designated drug conveyance, hyperthermia treatment, and attractive reverberation based diagnostics. The test lies in creating biocompatible attractive materials with exact command over their connections with natural frameworks.

4.1 Overview of different magnetic materials

Attractive materials assume a critical part in different mechanical applications, going from ordinary items like fridge magnets to cutting edge innovations, for example, attractive reverberation imaging (X-ray) machines and attractive stockpiling gadgets. These materials show attractive properties because of the arrangement of their nuclear or atomic attractive minutes. The investigation of attractive materials includes figuring out their characterization, properties, and applications. In this complete outline, we will investigate various sorts of attractive materials, their qualities, and their assorted applications.

Prologue to Attraction

Prior to digging into the points of interest of various attractive materials, understanding the fundamental standards of magnetism is fundamental. Attraction is a peculiarity that emerges from the movement of charged particles, especially electrons, inside molecules. The most essential unit of attraction is the attractive second, which is related with the characteristic rakish energy (turn) of electrons.

Materials can be arranged in view of their attractive way of behaving into three principal classes: ferromagnetic, antiferromagnetic, and ferrimagnetic. Moreover, materials can likewise show paramagnetic or diamagnetic way of behaving, contingent upon their reaction to an outside attractive field.

Ferromagnetic Materials

Ferromagnetic materials are portrayed by the presence of solid and extremely durable attractive minutes at the nuclear or sub-atomic level. These minutes will generally adjust lined up with one another, subsequent in plainly visible polarization even without an outer attractive field. The most well-known illustration of a ferromagnetic material is iron.

Iron and Steel

Iron is a notable ferromagnetic material with a high attractive penetrability. In its unadulterated structure, iron has a body-focused cubic gem structure, and its attractive minutes will more often than not adjust unexpectedly, making major areas of strength for a field. Steel, a compound of iron, is another broadly utilized ferromagnetic material. The expansion of carbon to press in steel improves its mechanical properties and can likewise impact its attractive way of behaving.

Extremely durable Magnets

Past iron and steel, different composites and mixtures are utilized to make extremely durable magnets with improved attractive properties. Alnico (aluminum, nickel, cobalt), samarium-cobalt, and neodymium-iron-boron are instances of materials utilized to create strong extremely durable magnets utilized in applications like electric engines, speakers, and attractive sensors.

Antiferromagnetic Materials

Antiferromagnetic materials display attractive minutes that will quite often adjust antiparallel to one another. This arrangement brings about an undoing of the perceptible attractive second, causing these materials to seem nonmagnetic without any an outside attractive field. Be that as it may, applying an outer field can initiate an attractive reaction.

Manganese Oxide

Manganese oxide (MnO) is an exemplary illustration of an antiferromagnetic material. In its antiferromagnetic state, neighboring manganese particles have antiparallel twists, prompting a net attractive snapshot of nothing. This kind of material is essential in the advancement of specific attractive sensors and spintronic gadgets.

Ferrimagnetic Materials

Ferrimagnetic materials are like antiferromagnets in that they have contradicting attractive minutes, yet in ferrimagnets, the extents of the minutes are inconsistent, bringing about a net attractive second. This inconsistent plan emerges from the presence of two sublattices with various attractive minutes.

Ferrites

Ferrites, like magnetite (Fe_3O_4), are significant ferrimagnetic materials. Magnetite comprises of both Fe^{2+} and Fe^{3+} particles with various attractive minutes. This outcomes in a net charge even without a trace of an outside field. Ferrites track down applications in transformers, inductors, and microwave gadgets.

Paramagnetic Materials

Paramagnetic materials have individual attractive minutes that will generally line up with an outer attractive field yet don't show unconstrained charge in that frame of mind of such a field. These materials have unpaired electrons, which add to their attractive powerlessness.

Aluminum

Aluminum is a typical paramagnetic material because of the presence of unpaired electrons in its external orbitals. While the attractive helplessness of paramagnetic materials is feeble, it can in any case be tackled in different applications, including attractive reverberation imaging (X-ray) in the clinical field.

Gadolinium

Gadolinium is one more paramagnetic component generally utilized in clinical imaging. Gadolinium-based contrast specialists are utilized to upgrade the perceivability of specific tissues in X-ray checks, gaining by the component's paramagnetic properties.

Diamagnetic Materials

Diamagnetic materials, rather than paramagnetic materials, have no unpaired electrons, and their attractive minutes will generally go against an applied attractive field. Subsequently, diamagnetic materials are pitifully repulsed by attractive fields.

Bismuth

Bismuth is an exemplary illustration of a diamagnetic material. Its electronic setup guarantees that all electrons are matched, prompting an absence of natural attractive minutes. Bismuth is much of the time utilized in examinations to show the aversion of diamagnetic materials from attractive fields.

Uses of Attractive Materials

Understanding the different properties of attractive materials permits specialists and researchers to foster many applications across different businesses.

Gadgets and Data Stockpiling

In the field of gadgets, attractive materials are critical for information capacity. Hard circle drives (HDDs) use ferromagnetic materials to store twofold information through the direction of attractive spaces. The advancement of elite execution magnets, for example, neodymium-iron-boron magnets, has likewise reformed the scaling down of electronic gadgets.

Clinical Applications

Attractive materials assume an essential part in clinical diagnostics and medicines. Attractive reverberation imaging (X-ray) depends on the attractive properties of specific cores, like hydrogen, inside the body. The utilization of paramagnetic differentiation specialists upgrades the difference in X-ray pictures, supporting the conclusion of different ailments.

Energy Change and Capacity

Attractive materials are necessary to the age and change of electrical energy. Super durable magnets are fundamental parts in electric generators, while attractive materials are utilized in transformers to manage voltage levels in power dispersion frameworks. Research is progressing to foster attractive materials for more proficient energy stockpiling in applications like battery-powered batteries.

Sensors and Actuators

Attractive sensors are broadly utilized in different applications, including route frameworks, auto control frameworks, and modern mechanization. Corridor impact sensors, for example, depend on the attractive field-prompted voltage across a semiconductor to distinguish the presence of an attractive field. Moreover, attractive materials are utilized in actuators, like solenoids and attractive valves, to control mechanical developments.

Ongoing Advances and Future Bearings

The field of attractive materials keeps on developing with continuous examination and innovative progressions. Late advancements incorporate the investigation of new attractive materials with upgraded properties, like higher attractive penetrability and further developed steadiness at outrageous temperatures. The coordination of attractive materials into arising advances like spintronics holds guarantee for novel applications in data handling and stockpiling.

Spintronics

Spintronics, short for turn transport gadgets, is an interdisciplinary field that takes advantage of the inborn twist of electrons for data handling and stockpiling. Attractive materials assume an essential part in spintronics, empowering the control and identification of electron turns. Spintronic gadgets, like attractive passage intersections, offer possible answers for non-unpredictable memory and high level figuring designs.

Attractive Nanomaterials

The advancement of attractive nanomaterials has opened up additional opportunities in medication, gadgets, and energy. Nanoscale attractive particles, for example, iron oxide nanoparticles, display interesting attractive properties that contrast from their mass partners. These materials track down applications in designated drug conveyance, hyperthermia disease medicines, and attractive reverberation imaging with further developed contrast.

4.2 Ferrimagnetic and antiferromagnetic substances

Attraction, a peculiarity established in the arrangement of nuclear and subatomic attractive minutes, appears in different materials with particular ways of behaving. Ferromagnetism, the most natural sort, includes equal arrangement of attractive minutes, bringing about areas of strength for a long-lasting perceptible polarization. Ferrimagnetism and antiferromagnetism, two less ordinarily known types, present interesting game plans of attractive minutes, prompting captivating properties and applications.

Ferrimagnetism: A Dualistic Arrangement

Ferrimagnetic materials, exemplified by intensifies like magnetite (Fe3O4), feature a dualistic arrangement of attractive minutes inside their precious stone designs. In contrast to ferromagnetic materials, ferrimagnets comprise of two attractive sublattices with inconsistent attractive minutes pointing in inverse headings. This unevenness creates a net plainly visible polarization, but less extraordinary than in ferromagnetic substances. A quintessential illustration of ferrimagnetism, magnetite owes its attractive way of behaving to the conjunction of Fe2+ and Fe3+ particles possessing unmistakable crystallographic destinations. The inconsistent extents and restricting directions of their attractive minutes bring about a net polarization, making magnetite an important material in different mechanical applications.

Ferrimagnetic materials are portrayed by their inconsistent attractive minutes, which make a net polarization. The crystallographic construction of these substances frequently includes two distinct kinds of attractive particles. A vital highlight note is the temperature reliance of ferrimagnetism. Like ferromagnetism, ferrimagnetic request is lost over a basic temperature known as the Curie temperature (Tc). This progress to a paramagnetic state highlights the meaning of temperature control in tackling the attractive properties of ferrimagnetic materials.

Uses of ferrimagnetic substances range different fields. In attractive capacity, ferrimagnets add to the advancement of hard drives and attractive tapes. The exceptional attractive properties of ferrimagnetic materials find application in microwave gadgets like isolators and circulators. Furthermore, ferrimagnetic sensors assume a critical part in different mechanical frameworks.

Antiferromagnetism: The Dropping Impact

Antiferromagnetic materials, as opposed to ferromagnets and ferrimagnets, show an antiparallel arrangement of attractive minutes. Nearby minutes point in inverse headings, bringing about a dropping impact on the perceptible scale. Manganese oxide (MnO) is an exemplary illustration of an antiferromagnetic substance. The gem construction of MnO directs a particular plan of manganese particles, where adjoining attractive minutes offset one another, yielding a material with zero naturally visible charge.

The most particular element of antiferromagnetic materials is the antiparallel arrangement of neighboring attractive minutes. This prompts a dropping impact, bringing about zero plainly visible charge. The basic temperature for antiferromagnetic materials, known as the Neel temperature (TN), denotes the progress to a paramagnetic state. Above TN, the antiferromagnetic request is lost, and the material acts as a paramagnet.

While antiferromagnetic materials may not be regularly utilized to make super durable magnets, they track down applications in arising advances. Spintronics, which uses the twist of electrons notwithstanding their charge, benefits from the novel properties of antiferromagnetic materials. Besides, antiferromagnetic substances assume a urgent part in magneto-optical gadgets, adding to headways in data handling and stockpiling.

Looking at Ferrimagnetism and Antiferromagnetism

A relative examination of ferrimagnetism and antiferromagnetism uncovers unmistakable contrasts in their attractive ways of behaving and applications.

In ferrimagnetic materials, the inconsistent attractive snapshots of two sublattices lead to a net plainly visible polarization. This stands as opposed to antiferromagnetic materials, where the antiparallel arrangement of neighboring minutes brings about zero net charge. Moreover, ferrimagnets frequently include gem structures with two distinct kinds of attractive particles, while antiferromagnets display explicit courses of action working with antiparallel arrangement.

The uses of these attractive ways of behaving additionally contrast. Ferrimagnetic materials, with their net charge, find utility in attractive capacity gadgets like hard drives. They are additionally urgent in the improvement of microwave gadgets, for example, isolators and circulators. Then again, antiferromagnetic materials add to the field of spintronics, where their remarkable properties improve the effectiveness and capacity limit of non-unstable memory. In addition, antiferromagnets are fundamental to the progression of magneto-optical gadgets, extending the skylines of data handling advances.

Mechanical Uses of Ferrimagnetic and Antiferromagnetic Materials

Attractive Capacity:

Ferrimagnetic materials, inferable from their net plainly visible charge, assume an essential part in attractive capacity gadgets. Hard drives, an omnipresent part in present day figuring, depend on the capacity of ferrimagnets to hold attractive data after some time. The remarkable attractive properties of ferrimagnetic materials add to the high-thickness capacity and quick recovery of information.

Microwave Gadgets:

The utilization of ferrimagnetic materials stretches out to the domain of microwave gadgets. Isolators and circulators, fundamental parts in microwave correspondence frameworks, influence the remarkable attractive attributes of ferrimagnets. These gadgets work with the unidirectional transmission of signs and safeguard touchy electronic parts from undesirable reflections.

Sensors:

Ferrimagnetic sensors track down application in different mechanical frameworks. The capacity of ferrimagnetic materials to display a quantifiable reaction to outer attractive fields makes them important in sensor advancements. From car applications to clinical gadgets, ferrimagnetic sensors add to the accuracy and dependability of different detecting instruments.

Spintronics:

Antiferromagnetic materials are at the very front of spintronics, a field that saddles the twist of electrons for data handling. Not at all like customary gadgets in view of electron charge, spintronics depends on the natural twist of electrons. Antiferromagnetic materials, with their antiparallel arrangement of attractive minutes, give a stage to controlling electron turns, prompting progressions in non-unpredictable memory and figuring innovations.

Magneto-Optical Gadgets:

Antiferromagnetic materials assume a vital part in magneto-optical gadgets, where the communication among attractive and optical properties is taken advantage of for data handling and stockpiling. The dropping impact of antiparallel attractive minutes is saddled to accomplish exact command over the polarization of light. This capacity tracks down applications in information capacity and correspondence advances.

Difficulties and Future Headings

While ferrimagnetic and antiferromagnetic materials offer exceptional attractive properties for different applications, challenges continue bridling their maximum capacity. One test lies in accomplishing and keeping the ideal attractive control at viable temperatures. The reliance of these materials on unambiguous precious stone designs and controlled conditions presents producing provokes that should be tended to for far and wide mechanical execution.

In the domain of antiferromagnetism, understanding and controlling the elements of attractive minutes at the nanoscale present continuous difficulties. Accomplishing stable antiferromagnetic request at surrounding temperatures is significant for pragmatic applications in spintronics and magneto-optical gadgets.

The future of ferrimagnetic and antiferromagnetic materials in innovation depends on progressions in material plan, fabricating processes, and a more profound comprehension of the hidden physical science. Specialists keep on investigating novel mixtures and nanoscale designs to push the limits of attractive materials for cutting edge applications.

4.3 Magnetic polymers and their emerging applications

Attractive polymers, an intriguing crossing point of polymer science and attraction, have arisen as a promising class of materials with different applications. These polymers consolidate the flexible properties of polymers with the one of a kind attractive qualities, opening new roads in fields like medication, gadgets, and natural remediation. In this thorough investigation, we will dive into the central ideas, amalgamation strategies, and arising uses of attractive polymers.

Figuring out Attractive Polymers

Polymer Rudiments

Polymers are enormous atoms made out of rehashing underlying units known as monomers. They show a great many physical and synthetic properties, making them crucial in different businesses. The adaptability in planning polymers permits scientists to fit their properties to meet explicit necessities, going from mechanical solidarity to biocompatibility.

Prologue to Attractive Polymers

Attractive polymers, as the name recommends, coordinate attractive functionalities into polymeric designs. This mix is accomplished by consolidating attractive nanoparticles, for example, iron oxide or ferrite nanoparticles, into the polymer grid. The subsequent materials hold the innate properties of the two polymers and attractive nanoparticles, offering a synergistic blend that opens up clever conceivable outcomes in material plan.

Blend Techniques

1. **In Situ Polymerization**

   One normal technique for blending attractive polymers is through in situ polymerization, where the polymerization cycle happens simultaneously with the fuse of attractive nanoparticles. This technique takes into account exact command over the scattering of attractive particles inside the polymer network, guaranteeing homogeneity and advancing attractive properties.

2. **Arrangement Mixing**

   In arrangement mixing, pre-blended polymers and attractive nanoparticles are consolidated in an answer. The blend is then handled to frame the last attractive polymer composite. This technique offers adaptability in picking polymers with explicit properties and changing the nanoparticle stacking for wanted attractive way of behaving.

3. **Emulsion Polymerization**

   Emulsion polymerization includes scattering monomers and attractive nanoparticles in an emulsion. The polymerization cycle happens in the watery stage, prompting the development of attractive polymer nanoparticles. This strategy is especially valuable for creating attractive polymer microspheres with controlled sizes and attractive properties.

4. **Soften Blending**

Soften blending is a direct strategy where attractive nanoparticles are truly blended in with liquid polymers. The blend is then set to frame the attractive polymer composite. While this strategy is somewhat basic, it might present difficulties in accomplishing uniform dispersion and ideal attractive properties.

Attractive Properties of Attractive Polymers

The attractive properties of attractive polymers are impacted by a few variables, including the sort and size of attractive nanoparticles, their conveyance inside the polymer lattice, and the general composite design. Generally utilized attractive nanoparticles incorporate iron oxide (Fe_3O_4) and ferrites, picked for their attractive steadiness and biocompatibility.

1. **Polarization**

   Attractive polymers show charge conduct like that of attractive nanoparticles. The presence of attractive spaces inside the polymer framework permits the material to answer outside attractive fields, prompting applications in regions like attractive detecting and activation.

2. **Superparamagnetism**

   In nanoscale attractive particles, superparamagnetism becomes conspicuous. This peculiarity, described by the shortfall of attractive hysteresis, takes into consideration effective reaction to outside attractive fields and forestalls remaining charge, making attractive polymers reasonable for applications requiring reversible polarization changes.

3. **Tunable Attractive Properties**

The attractive properties of attractive polymers can be tweaked by changing elements, for example, the nanoparticle focus, molecule size, and the polymer network's temperament. This tunability upgrades the adaptability of attractive polymers, empowering their customization for explicit applications.

Arising Uses of Attractive Polymers

The one of a kind mix of polymer adaptability and attractive functionalities has pushed attractive polymers into a domain of different applications. Analysts and enterprises are investigating these materials for their likely in different fields.

1. **Biomedical Applications**

1. **Attractive Medication Conveyance**

   One of the most encouraging uses of attractive polymers in medication is in the field of medication conveyance. Attractive polymers can be intended to epitomize tranquilizes and convey them to explicit objective destinations inside the body affected by an outside attractive field. This designated drug conveyance approach limits aftereffects and upgrades the remedial adequacy of medications.

2. **Attractive Hyperthermia**

   Attractive hyperthermia includes involving attractive polymers for confined warming of tissues. When presented to a substituting attractive field, attractive nanoparticles inside the polymer framework create heat through hysteresis misfortunes. This intensity can be used to specifically obliterate malignant growth cells in hyperthermia-based disease treatments.

3. **Attractive Reverberation Imaging (X-ray) Difference Specialists**

Attractive polymers have found application as difference specialists in attractive reverberation imaging (X-ray). The attractive nanoparticles implanted in the polymer framework upgrade the differentiation in imaging, giving more clear and more itemized pictures for demonstrative purposes.

2. Gadgets and Data Stockpiling

1. **Attractive Sensors**

   The polarization conduct of attractive polymers makes them reasonable for use in attractive sensors. These sensors track down applications in different electronic gadgets, including route frameworks, auto sensors, and modern robotization.

2. **Information Capacity**

Attractive polymers add to progressions in information capacity advancements. The incorporation of attractive nanoparticles into polymeric lattices considers the advancement of adaptable and lightweight attractive stockpiling materials, possibly changing the field of data stockpiling.

3. Natural Remediation

1. **Water Purging**

   Attractive polymers can be utilized for the expulsion of pollutants from water. The attractive properties work with the simple partition of the polymer composite from water subsequent to adsorbing poisons, empowering effective and reusable water purging frameworks.

2. **Oil slick Cleanup**

In natural crises, for example, oil slicks, attractive polymers can be conveyed for effective oil-water partition. The attractive idea of the polymers empowers the assortment of oil, working on cleanup cycles and decreasing the natural effect of such episodes.

4. Adaptable Gadgets

Attractive polymers, with their mix of mechanical adaptability and attractive usefulness, hold guarantee in the field of adaptable gadgets. These materials can be integrated into adaptable electronic gadgets, like bendable showcases and wearable hardware, extending the opportunities for cutting edge innovations.

5. Brilliant Materials and Actuators

The responsiveness of attractive polymers to outside attractive fields makes them appropriate for use in savvy materials and actuators. These materials can go through controlled shape changes or mechanical reactions affected by an attractive field, opening roads for imaginative applications in advanced mechanics and bio-medical gadgets.

6. Energy Reaping

The utilization of attractive polymers in energy collecting has acquired consideration. The capacity of these materials to change over mechanical vibrations into electrical energy, known as magnetostriction, presents valuable open doors for creating energy-reaping gadgets that can control little electronic parts.

Difficulties and Future Possibilities

While attractive polymers hold huge commitment, a few difficulties should be tended to for their far reaching reception.

1. **Amalgamation Control and Scale-up**

   Accomplishing exact command over the union of attractive polymers, particularly as far as nanoparticle scattering and size dissemination, stays a test. Also, increasing creation processes while keeping up with the ideal properties presents troubles that should be defeated for modern applications.

2. **Biocompatibility and Harmfulness**

   In biomedical applications, guaranteeing the biocompatibility of attractive polymers is basic. Specialists should completely explore potential harmfulness issues and long haul impacts to ensure the wellbeing of these materials for use inside the human body.

3. **Strength and Solidness**

The strength and solidness of attractive polymers in different conditions, particularly those including mechanical pressure or temperature changes, should be totally assessed. Guaranteeing the drawn out usefulness of these materials is significant for their progress in applications like hardware and ecological remediation.

4. **Cost Contemplations**

The expense of delivering attractive polymers, especially those expected for huge scope applications, is an element that needs cautious thought. Finding financially savvy combination techniques and upgrading material use will be vital to making these materials economically feasible.

5. **Administrative Endorsements**

For biomedical applications, administrative endorsements are fundamental. Attractive polymers expected for use in drug conveyance or clinical imaging should go through thorough testing to fulfill security and adequacy guidelines prior to arriving at the market.

Looking forward, the fate of attractive polymers is promising. Proceeded with research endeavors pointed toward addressing these difficulties will probably prompt leap forwards, making ready for the boundless reception of attractive polymers across different enterprises.

Chapter 5

Technological Applications Of Iron's Magnetism

Iron, a pervasive component with attractive properties, has been a foundation in the improvement of different advances that have changed the cutting edge world. Its exceptional attractive attributes, particularly as ferromagnetism, have been outfit across different fields, going from gadgets and broadcast communications to clinical diagnostics and energy creation. In this exhaustive investigation, we will dive into the mechanical utilizations of iron's attraction, revealing its essential job in molding the innovative scene.

Grasping Iron's Attraction

Prior to digging into explicit applications, understanding the attractive properties of iron is fundamental. At the nuclear and sub-atomic levels, iron has attractive minutes because of the arrangement of its electrons. In mass materials, especially in the translucent construction of iron, the arrangement of these attractive minutes leads to ferromagnetism. Ferromagnetic materials, for example, iron, show an unconstrained polarization where adjoining attractive minutes adjust lined up with one another, subsequent in a solid and diligent attractive field.

Iron in Gadgets and Broadcast communications

1. **Electromagnetic Enlistment and Transformers**

   Iron's attraction is major to the activity of transformers, gadgets that assume an essential part in electrical power conveyance. While an exchanging current (AC) moves through a curl twisted around an iron center, the changing attractive field prompts a voltage in a close by loop. This rule of electromagnetic enlistment is outfit in transformers to move forward or step down voltage levels, empowering effective transmission and conveyance of electrical energy across power lattices.

2. **Electromagnets**

   The utilization of iron in electromagnets is a foundation in different mechanical applications. By folding a loop of wire over an iron center and passing an electric flow through the curl, the iron becomes charged, making major areas of strength for a field. Electromagnets track down applications in electric engines, speakers, attractive locks, and attractive reverberation imaging (X-ray) machines in the clinical field.

3. **Attractive Capacity Gadgets**

   Iron assumes a significant part in attractive capacity gadgets, reforming the manner in which we store and access data. Hard plate drives (HDDs) utilize ferromagnetic materials, including iron, to store parallel information through the direction of attractive areas. Each piece of data compares to the arrangement of attractive minutes in the material. The capacity of iron to hold charge makes it a fundamental part in the capacity innovation that underlies present day PCs.

4. **Attractive Sensors and Actuators**

Iron-based attractive sensors are essential parts in a horde of utilizations, including compasses, attractive field sensors in cell phones, and auto sensors. The Lobby impact, a peculiarity where a voltage is created opposite to both the current and attractive field in a guide, is taken advantage of in Corridor impact sensors. Furthermore, iron's attraction adds to the activity of attractive actuators, for example, solenoids, which are fundamental in different mechanical frameworks.

Iron in Clinical Diagnostics

1. **Attractive Reverberation Imaging (X-ray)**

   One of the most groundbreaking utilizations of iron's attraction in the clinical field is in Attractive Reverberation Imaging (X-ray). In X-ray, major areas of strength for machines fields produced by superconducting magnets adjust the attractive snapshots of hydrogen iotas in the body. Radiofrequency beats are then used to bother this arrangement, and the ensuing unwinding processes radiate transmissions that are changed over into point by point, high-goal pictures of inward body structures. Iron-based contrast specialists are now and then used to improve imaging in unambiguous cases.

2. **Attractive Molecule Imaging (MPI)**

   Attractive Molecule Imaging is an arising imaging strategy that depends on iron oxide nanoparticles as tracers. These nanoparticles, when infused into the body, answer an outside attractive field, delivering signals that are utilized to make continuous, high-goal pictures. MPI shows guarantee as a harmless imaging methodology with expected applications in cardiovascular imaging, disease identification, and other clinical diagnostics.

3. **Drug Conveyance and Hyperthermia**

Iron's attractive properties are utilized in the field of nanomedicine for designated drug conveyance and hyperthermia. Nanoparticles made of iron oxide can be functionalized with drugs and coordinated to explicit locales in the body utilizing an outside attractive field. This designated drug conveyance limits aftereffects and improves the restorative adequacy of prescriptions. Moreover, iron-based nanoparticles can create heat when presented to rotating attractive fields, a peculiarity used in hyperthermia therapies for disease.

Iron in Natural and Energy Applications

1. **Natural Remediation**

   Iron's attraction assumes an essential part in natural remediation endeavors, especially in the expulsion of toxins from water and soil. Iron nanoparticles can be utilized to adsorb and immobilize contaminations, working with their expulsion from the climate. The attractive properties of these nanoparticles empower simple division from the treated water or soil, making the remediation cycle more effective.

2. **Attractive Levitation (Maglev) Trains**

   In transportation, iron's attractive properties find application in Maglev prepares, a fast rail innovation. Maglev trains utilize strong magnets, frequently made with iron or iron-based materials, to suspend over the tracks and push the train forward. This frictionless method of transportation offers high paces and energy proficiency, displaying the groundbreaking capability of iron's attraction in the domain of current travel.

3. **Energy Age and Capacity**

Iron-based materials add to energy age and capacity innovations. In the domain of sustainable power, iron oxide nanoparticles are researched for their true capacity in sun based cell applications. Furthermore, iron-based materials are utilized in battery-powered batteries, with continuous exploration zeroed in on upgrading the presentation and life expectancy of these batteries for boundless use in electric vehicles and sustainable power frameworks.

Iron in Development and Foundation

1. **Attractive Cement**

   Advancements in development materials have prompted the improvement of attractive cement, which integrates iron particles in with the general mishmash. This attractive cement can be controlled with an outer attractive field, offering additional opportunities in development and framework projects. Applications incorporate the production of self-recuperating structures and the potential for attractively controlled development robots.

2. **Attractive Entryway Seals and Hostile to Vibration Frameworks**

Iron's attractive properties add to the improvement of attractive entryway seals that give areas of strength for a powerful seal against outside components. What's more, attractive enemy of vibration frameworks use iron-based materials to hose vibrations in hardware and designs, diminishing commotion and improving underlying honesty.

Difficulties and Future Bearings

While iron's attraction has energized various innovative headways, there are difficulties and progressing research regions that warrant consideration.

1. **Material Advancement**

   Advancing the properties of iron-based materials, for example, attractive nanoparticles, is a continuous test. Specialists are investigating ways of upgrading the attractive attributes, security, and biocompatibility of these materials for further developed execution in applications like clinical imaging and medication conveyance.

2. **Energy Proficiency**

   Endeavors to further develop the energy proficiency of advances using iron's attraction are fundamental. This remembers growing more productive attractive materials for use for engines, transformers, and different gadgets, as well as investigating novel ways to deal with energy age and capacity.

3. **Biocompatibility and Security**

   In clinical applications, guaranteeing the biocompatibility and security of iron-based nanoparticles is urgent. Understanding the expected long haul impacts of these materials inside the human body is an area of progressing research, particularly as nanomedicine keeps on progressing.

4. **Practical Union Strategies**

   Creating practical union strategies for iron-based materials is a thought for far reaching reception in different applications. The financial suitability of advances, for example, attractive capacity gadgets and clinical imaging depends on effective and versatile assembling processes.

5. **Investigation of New Applications**

Proceeded with investigation of new applications for iron's attraction is a unique area of exploration. As how we might interpret the material develops, there is potential for novel advances and developments that can additionally coordinate iron's attractive properties into different fields.

5.1Electromagnetism and its role in technology

Electromagnetism, a principal power of nature, has been tackled and formed by humankind to control innumerable mechanical developments. This complex power, emerging from the association between electric charges and attractive fields, frames the premise of electromagnetism. From the creation of electric engines and generators to the improvement of trend setting innovations like attractive reverberation imaging (X-ray) and remote correspondence, electromagnetism has been at the front line of mechanical advancement. In this investigation, we will dive into the standards of electromagnetism, its verifiable importance, and its groundbreaking job across different mechanical spaces.

Figuring out Electromagnetism

1. **The Central Powers**

   At the core of electromagnetism lies the perplexing connection between electric charges and attractive fields. This power, one of the four central powers of nature, binds together electric and attractive peculiarities into a strong structure known as electromagnetism. The electromagnetic power oversees the way of behaving of charged particles, affecting their movement and collaborations.

2. **Maxwell's Conditions**

The numerical detailing of electromagnetism is carefully caught by Maxwell's situations, a bunch of four conditions created by James Representative Maxwell in the nineteenth hundred years. These conditions portray how electric and attractive fields connect and proliferate through space, giving an extensive system to grasping the way of behaving of electromagnetic waves. Maxwell's noteworthy work established the groundwork for the advanced comprehension of electromagnetism and made ready for mechanical forward leaps.

Authentic Importance

1. **Revelation of Electromagnetic Acceptance**

   The nineteenth century saw critical advancements in electromagnetism, quite the disclosure of electromagnetic enlistment. Michael Faraday, in a progression of momentous examinations, showed the way that a changing attractive field could prompt an electric flow in a close by conveyor. This disclosure, known as Faraday's law of electromagnetic acceptance, laid the preparation for the advancement of electric generators, it is created and disseminated to change the way power.

2. **Electromagnetic Message**

   The mid-nineteenth century saw the commonsense use of electromagnetism in correspondence with the creation of the electromagnetic message. Samuel Morse and Alfred Vail fostered a framework that pre-owned electromagnets to control the opening and shutting of circuits, permitting the transmission of coded messages over significant distances. This creation altered significant distance correspondence, denoting the appearance of current media transmission frameworks.

3. Electromagnetic Waves and Remote Correspondence

Expanding upon Maxwell's situations, Heinrich Hertz tentatively affirmed the presence of electromagnetic waves in the late nineteenth hundred years. These waves, presently known as radio waves, structure the premise of remote correspondence. The improvement of radio innovation, led by innovators like Guglielmo Marconi, took into consideration the transmission of data over significant distances without the requirement for actual associations. This established the groundwork for present day media communications and made ready for developments like TV and versatile correspondence.

Electromagnetism in Innovation

1. **Electric Engines**

   One of the earliest and most extraordinary utilizations of electromagnetism is found in electric engines. The standard of electromagnetic enlistment, combined with the connection between attractive fields and electric flows, empowers the transformation of electrical energy into mechanical energy. Electric engines power a huge swath of gadgets, from home devices to modern hardware, driving the wheels of innovative advancement.

2. **Generators and Power Age**

   Equally, the guideline of electromagnetic enlistment is at the center of electric generators. By pivoting a curl inside an attractive field, generators convert mechanical energy into electrical energy. This cycle is central to control age in customary power plants, where the rotational movement is frequently determined by steam turbines or different wellsprings of mechanical energy. The capacity to create power for a huge scope has been instrumental in fueling social orders and cultivating mechanical progressions.

3. **Attractive Reverberation Imaging (X-ray)**

   In the domain of clinical diagnostics, electromagnetism assumes a pivotal part in the improvement of attractive reverberation imaging (X-ray). This exceptional imaging method uses solid attractive fields and radiofrequency heartbeats to create nitty gritty pictures of inward designs inside the human body. The capacity to imagine delicate tissues with high accuracy has changed clinical analysis and treatment arranging.

4. **Electromagnetic Sensors and Finders**

   Electromagnetic sensors and finders are unavoidable in different mechanical applications. From the compass, which uses the World's attractive field for route, to refined sensors in security frameworks and modern gear, the capacity to distinguish and quantify electromagnetic fields has assorted pragmatic applications.

5. **Electromagnetic Levitation**

   Maglev (attractive levitation) innovation is a striking illustration of the extraordinary effect of electromagnetism on transportation. Maglev trains, impelled by the repugnance and fascination of attractive fields, drift above tracks without actual contact, limiting erosion and considering high velocity travel. This innovation can possibly reform the eventual fate of transportation with its effectiveness and speed.

6. **Remote Charging**

   The idea of remote charging, made conceivable by electromagnetism, has acquired noticeable quality lately. Gadgets, for example, cell phones and electric vehicles can be charged without actual associations using electromagnetic enlistment. This innovation upgrades client accommodation and diminishes dependence on conventional wired charging techniques.

7. **Electromagnetic Warming and Enlistment Cooking**

   Electromagnetic warming, frequently utilized in enlistment cooking, is one more utilization of electromagnetism. In enlistment cooktops, an exchanging current goes through a curl underneath the cookware, producing a wavering attractive field. This attractive field actuates electric flows in the conductive cookware, prompting resistive warming and productive cooking.

8. **Molecule Gas pedals**

   In the domain of logical exploration, molecule gas pedals use electromagnetic fields to speed up charged particles to high rates. These strong machines, similar to the Huge Hadron Collider (LHC), empower physicists to concentrate on the crucial structure blocks of issue and unwind the secrets of the universe.

9. **Electromagnetic Locks**

Electromagnetic locks, ordinarily utilized in access control frameworks and security applications, depend on the guideline of electromagnetic fascination. At the point when an electric flow goes through a curl, it produces an attractive field that draws in a metal plate, making a protected locking component without the requirement for actual contact.

Difficulties and Future Bearings

While electromagnetism has impelled mechanical headways, there are difficulties and continuous exploration regions that merit consideration.

1. **Productivity and Energy Utilization**

   Productivity stays a key thought, particularly in electromechanical frameworks like electric engines and generators. Propels in materials and configuration are expected to upgrade effectiveness and diminish energy utilization, adding to the improvement of additional supportable advancements.

2. **Scaling down and Mix**

   As innovation keeps on developing, the interest for more modest and more coordinated gadgets increments. Accomplishing scaling down while keeping up with execution presents difficulties, requiring advancements in materials, producing cycles, and plan.

3. **Electromagnetic Similarity**

   The expansion of electronic gadgets presents difficulties connected with electromagnetic similarity (EMC). Guaranteeing that gadgets work without obstruction within the sight of electromagnetic fields is essential for the solid working of mind boggling mechanical frameworks.

4. **Propels in Attractive Materials**

   Proceeded with research in attractive materials is fundamental for growing more proficient gadgets and advances. The mission for novel materials with upgraded attractive properties, for example, high coercivity and low energy misfortunes, drives progress in regions like information stockpiling and energy transformation.

5. **Investigation of Quantum Electrodynamics**

At the very front of hypothetical physical science, the investigation of quantum electrodynamics (QED) expects to accommodate quantum mechanics with electromagnetism. Understanding the way of behaving of electromagnetic fields at the quantum level could prompt forward leaps in processing, correspondence, and material science.

5.2 Magnetic storage devices: from floppy disks to modern hard drives

Attractive capacity gadgets play had a crucial impact in the development of data stockpiling and recovery, filling in as the foundation of computerized information the executives for a very long time. From the unassuming starting points of floppy circles to the complex innovation of current hard drives, attractive capacity has gone through a momentous excursion, set apart by consistent development and mechanical progressions. In this complete investigation, we will dive into the authentic improvement of attractive stockpiling gadgets, analyzing the key achievements, basic advances, and the groundbreaking effect these gadgets have had on the manner in which we store and access information.

Early Days: The Ascent of Attractive Tape and Drum Stockpiling

1. **Attractive Tape Stockpiling**

   The commencement of attractive stockpiling can be followed back to the mid-twentieth 100 years, with the appearance of attractive tape stockpiling. Created in the mid 1950s, attractive tape frameworks used reels of tape covered with an attractive material. Data was put away as attractive examples on the tape, and early applications included information capacity for PCs and sound recording.

Attractive tape offered a financially savvy and proficient method for putting away huge volumes of information consecutively. It tracked down early reception in centralized server PCs for recorded purposes and information reinforcement. The consecutive access nature of tape stockpiling, where information is perused or composed successively, made it reasonable for applications requiring high-limit capacity yet not really quick access.

2. Attractive Drum Stockpiling

Simultaneously, attractive drum stockpiling arose as one more early type of attractive stockpiling innovation. Attractive drums comprised of a turning drum covered with an attractive material, and information was put away as polarized spots on the drum's surface. Attractive drum stockpiling was utilized in early PCs like the UNIVAC I, giving an arbitrary access capacity to quicker information recovery contrasted with attractive tape.

While both attractive tape and drum stockpiling were progressive for their time, they were restricted by their consecutive or semi-successive access nature and missing the mark on flexibility required for the intuitive registering conditions that would arise in the years to come.

The Time of Floppy Circles: Versatile Attractive Stockpiling

1. Presentation of Floppy Plates

The 1970s saw a critical jump in attractive capacity with the presentation of floppy circles. Created by IBM, the main floppy circles were 8-inch in distance across and housed in adaptable plastic sleeves. These early floppy plates had an attractive covering that put away information in a roundabout, concentric configuration. They were perused and composed utilizing a floppy plate drive (FDD) that highlighted a read/compose head to connect with the attractive surface.

Floppy plates addressed a change in outlook in attractive capacity because of their transportability and further developed irregular access capacities. They turned into the essential mechanism for PC clients to store and share information. As innovation progressed, floppy circles developed into more modest structure factors, with 5.25-inch and 3.5-inch plates becoming norm before long.

2. Floppy Plate Drives in PCs

The pervasiveness of floppy circle drives in PCs during the 1980s and 1990s significantly affected the openness and appropriation of programming and information. Floppy plates were utilized for information capacity as well as filled in as a method for booting working frameworks and running programming applications.

The "save" and "burden" activities became inseparable from embedding and shooting floppy plates, making them a basic piece of the figuring experience.

While floppy circles were progressive by their own doing, their restricted limit and helplessness to actual harm provoked the requirement for higher-limit and more vigorous attractive stockpiling arrangements.

Hard Plate Drives: The Transformation in Attractive Capacity

1. **Development of the Hard Plate Drive (HDD)**

   The coming of hard plate drives denoted a groundbreaking stage in the development of attractive stockpiling. The principal hard circle drive, the IBM 305 RAMAC, was presented in 1956. Not at all like floppy circles, hard drives included unbending, attractive platters that turned at high rates. Information was perused and composed utilizing attractive heads mounted on actuator arms that got across the platter surfaces.

   The IBM 305 RAMAC had a capacity limit of 5 megabytes and was basically utilized for business and logical applications. While it was enormous and costly by the present guidelines, the idea of a high-limit, irregular access stockpiling gadget established the groundwork for the fate of attractive stockpiling.

2. **Development of Hard Circle Drives**

   As innovation progressed, hard drives went through critical upgrades with regards to capacity limit, speed, and structure factor. The presentation of Winchester hard plate drives during the 1970s achieved a significant development - the joining of fixed units that kept a residue free climate for the drive's inward parts. This development laid the foundation for the cutting edge hard plate drives that would become indispensable to PCs.

   The 1980s and 1990s saw a quick expansion in hard drive limits and a decline in actual size. Hard drives changed from huge, independent units to smaller, inward drives that could fit inside work area and, ultimately, PCs. The advancement of more modest structure factors, including the 2.5-inch drives generally utilized in workstations, added to the conveyability and far reaching reception of PCs.

3. **Progressions in Attractive Recording Advancements**

   The constant mission for higher capacity densities prompted progressions in attractive recording advancements. Longitudinal recording, where attractive particles are adjusted lined up with the plate surface, was the prevailing technique for quite some time. In any case, as capacity densities moved toward hypothetical restricts, a shift to opposite attractive recording (PMR) happened.

   Opposite attractive recording, presented during the 2000s, considered higher areal densities by adjusting attractive particles opposite to the plate surface.

This advanced essentially expanded capacity limits, empowering hard drives to store terabytes of information.

4. Strong State Drives (SSDs): Another Time

While customary hard plate drives stay common, the rise of strong state drives (SSDs) has presented another time away innovation. SSDs use NAND-based streak memory, which needs moving parts and depends on electronic circuits to store and recover information. This takeoff from the mechanical parts of conventional hard drives offers a few benefits, including quicker read/compose speeds, lower power utilization, and more prominent toughness.

SSDs have become progressively well known in customer gadgets, including workstations and cell phones, because of their speed and unwavering quality. Notwithstanding, conventional hard plate drives keep on assuming an essential part in information capacity, especially for applications requiring huge limits at a lower cost for every gigabyte.

Attractive Capacity in the 21st Hundred years: Difficulties and Advancements

1. **Challenges in Scaling Attractive Capacity**

   The constant interest for higher capacity limits and quicker information access presents difficulties in scaling attractive capacity advances. Accomplishing higher areal densities, where more information can be put away in a given actual space, requires developments in materials, read/compose advances, and blunder rectification components.

2. **Heat-Helped Attractive Recording (HAMR)**

   Heat-helped attractive recording (HAMR) is a promising innovation intended to defeat the restrictions of conventional attractive recording techniques. In HAMR, a laser is utilized to warm the circle surface prior to composing information, making it simpler to polarize the recording media. This approach considers higher capacity densities, pushing the limits of attractive stockpiling limits.

3. **Shingled Attractive Recording (SMR)**

   Shingled attractive recording (SMR) is one more development in attractive capacity design. In SMR drives, information tracks cross-over like shingles on a rooftop, empowering higher areal densities. While SMR offers expanded capacity limits, it presents difficulties as far as compose execution because of the need to change adjoining tracks while refreshing information.

4. **Mixture Capacity Arrangements**

   Mixture capacity arrangements, joining customary hard circle drives with a more modest SSD reserve, intend to find some kind of harmony among limit and speed. This approach use the huge limits of hard drives for mass capacity while using the quicker perused/compose rates of SSDs for much of the time got to information. Half breed drives give a financially savvy split the difference to clients looking for both more than adequate extra room and further developed execution.

5. **Distributed storage and Attractive Tapes**

Notwithstanding neighborhood stockpiling arrangements, the 21st century has seen the ascent of distributed storage, where information is put away and gotten to over the web. While not stringently attractive in the conventional sense, distributed storage frequently depends on monstrous server farms furnished with attractive tape libraries for documented purposes. Attractive tapes, a respected innovation, offer practical long haul stockpiling with high limits, making them reasonable for reinforcement and recorded applications.

5.3 Magnetic resonance imaging (MRI) and its impact on medical diagnostics

In the domain of clinical diagnostics, Attractive Reverberation Imaging (X-ray) remains as a progressive and harmless imaging strategy that has changed the scene of medical care. Since its beginning during the 1970s, X-ray has developed into a modern and broadly used instrument for imagining interior designs of the human body. Its capacity to give definite and high-goal pictures without the utilization of ionizing radiation has made it a foundation in the conclusion and checking of a heap of ailments. In this exhaustive investigation, we will dive into the standards of X-ray, its authentic turn of events, the mechanical headways that have formed it, and its significant effect on clinical diagnostics.

Understanding the Standards of Attractive Reverberation Imaging

1. **Rudiments of Atomic Attractive Reverberation (NMR)**

   The groundwork of X-ray lies in the standards of Atomic Attractive Reverberation (NMR), a peculiarity saw when certain nuclear cores collaborate with an outside attractive field. With regards to X-ray, the hydrogen cores (protons) in water atoms are fundamentally focused on. When presented to serious areas of strength for a field, these protons fall in line with the attractive field.

2. **Radiofrequency Heartbeats and Reverberation**

   To create pictures, radiofrequency beats are applied, briefly disturbing the arrangement of the protons. As the protons return to their adjusted state, they emanate radiofrequency signals. The recognition and investigation of these signs permit the development of point by point pictures. The expression "reverberation" in X-ray alludes to the particular recurrence at which protons retain and emanate energy when presented to a radiofrequency beat.

3. **Tissue Difference in X-ray**

The changing convergences of water and fat in various tissues impact the X-ray signal and add to tissue contrast. Boundaries like T1 (turn cross section unwinding time) and T2 (turn unwinding time) assume essential parts in deciding the difference in X-ray pictures. By controlling these boundaries, radiologists can acquire pictures that feature explicit tissues or neurotic circumstances.

Authentic Advancement of X-ray

1. **Rise of Atomic Attractive Reverberation (NMR)**

   The underlying foundations of X-ray can be followed back to the mid-twentieth century when researchers were investigating the peculiarity of Atomic Attractive Reverberation (NMR) for logical purposes. During the 1940s and 1950s, analysts like Felix Bloch and Edward Purcell autonomously found NMR, for which they were granted the Nobel Prize in Physical science in 1952.

2. **First X-ray Investigation**

   The main X-ray try was led in 1971 by Raymond Damadian, who showed the way that various tissues could create particular signs utilizing NMR. He followed this achievement with the improvement of the primary human X-ray scanner, which he named the "Unyielding." In 1977, the principal entire body X-ray check was performed on a human, denoting a vital crossroads throughout the entire existence of clinical imaging.

3. **Clinical Reception and Headways**

   All through the 1980s and 1990s, X-ray went through fast innovative progressions, including the advancement of more grounded magnets, further developed imaging arrangements, and the presentation of differentiation specialists. These advancements upgraded the quality and proficiency of X-ray filters. The clinical reception of X-ray extended, and it turned into an imperative device in different clinical strengths.

4. **Useful X-ray (fMRI) and Then some**

The late twentieth century saw the development of practical X-ray (fMRI), a strategy that actions changes in blood stream to recognize mind action. fMRI has become instrumental in neuroscience, permitting scientists and clinicians to plan cerebrum works and study conditions like epilepsy and mental problems.

Mechanical Parts of X-ray

1. **Magnet Frameworks**

   Fundamental to X-ray innovation is the magnet framework, which gives the strong and uniform attractive field essential for the imaging system. Early X-ray scanners utilized superconducting magnets cooled by fluid helium. In any case, progressions have prompted the advancement of high-temperature superconducting magnets and extremely durable magnets, decreasing the dependence on cryogenic cooling.

2. **Radiofrequency Loops**

Radiofrequency loops are fundamental parts that communicate the radiofrequency beats and get the transmissions discharged by protons. Various curls, including surface loops and staged exhibit loops, might be utilized to advance picture quality and goal for explicit locales of the body.

3. **Inclination Curls**

Slope curls make spatial varieties in the attractive field, empowering the restriction of signs from explicit areas. By controlling the slopes, X-ray scanners can create nitty gritty two-and three-layered pictures. Slope curls assume a significant part in deciding the cut thickness and picture goal.

4. **PC Frameworks and Programming**

The intricacy of X-ray information procurement and handling requires progressed PC frameworks and programming. Remaking calculations, picture handling programming, and UIs add to the proficiency and precision of X-ray assessments. Furthermore, computerized reasoning (artificial intelligence) and AI applications are progressively being incorporated into X-ray innovation for picture examination and translation.

5. **Contrast Specialists**

Contrast specialists, frequently founded on gadolinium, are utilized in X-ray to improve the perceivability of explicit tissues or anomalies. These specialists change the unwinding seasons of adjacent protons, bringing about expanded picture contrast. The advancement of novel differentiation specialists keeps on being an area of examination for working on demonstrative abilities.

Clinical Uses of X-ray

1. **Neuroimaging**

X-ray has turned into a foundation in neuroimaging, giving nitty gritty pictures of the cerebrum and spinal string. It is utilized to analyze and screen conditions like mind growths, strokes, different sclerosis, and neurodegenerative problems. Utilitarian X-ray (fMRI) is utilized for planning mind movement and network.

2. **Outer muscle Imaging**

In outer muscle imaging, X-ray is significant for surveying joints, tendons, ligaments, and delicate tissues. It helps with the determination and the executives of conditions like games wounds, joint inflammation, and outer muscle cancers. High-goal pictures empower muscular specialists to precisely design intercessions.

3. **Cardiovascular Imaging**

   Cardiovascular X-ray is used to picture the heart and veins, giving bits of knowledge into heart capability and distinguishing cardiovascular infections. It is especially significant for surveying inborn heart deserts, myocardial dead tissue, and cardiovascular growths.

4. **Stomach and Pelvic Imaging**

   X-ray assumes a vital part in stomach and pelvic imaging, offering nitty gritty perspectives on organs like the liver, kidneys, pancreas, and regenerative organs. It is utilized for diagnosing conditions like liver growths, kidney illnesses, and pelvic problems.

5. **Bosom Imaging**

   Bosom X-ray is utilized as a reciprocal device in bosom disease conclusion and screening. It is especially significant for assessing the degree of illness, surveying therapy reaction, and screening people at high gamble of bosom disease.

6. **Oncological Imaging**

   X-ray is generally utilized in oncological imaging for recognizing and organizing different diseases. It helps with portraying growths, surveying therapy reaction, and arranging medical procedures or radiation treatment. The utilization of useful imaging strategies, for example, dissemination weighted imaging, improves the capacity to recognize and screen growths.

7. **Pediatric Imaging**

   The non-ionizing nature of X-ray makes it appropriate for pediatric imaging. It is utilized to assess innate oddities, evaluate neurological circumstances, and analyze outer muscle problems in youngsters. Sedation might be utilized to guarantee the participation of youthful patients during the assessment.

8. **Interventional X-ray**

In interventional X-ray, continuous imaging is utilized to direct negligibly obtrusive systems. This incorporates biopsies, catheter positions, and cancer removals. The capacity to imagine the objective region continuously improves the accuracy and wellbeing of these intercessions.

Benefits of X-ray in Clinical Diagnostics

1. **Non-Ionizing Radiation**

   One of the vital benefits of X-ray is that it doesn't utilize ionizing radiation, in contrast to other imaging modalities like X-beams and registered tomography (CT). This makes X-ray especially appropriate for dreary or longitudinal investigations, lessening the potential dangers related with radiation openness.

2. **Prevalent Delicate Tissue Difference**

   X-ray gives prevalent delicate tissue contrast, taking into account point by point representation of organs, muscles, and other delicate tissues. This is particularly important in separating among typical and obsessive tissues and in recognizing irregularities that might be trying to relate to other imaging modalities.

3. **Multiplanar Imaging**

X-ray empowers multiplanar imaging, implying that pictures can be gotten in different planes (sagittal, coronal, and pivotal) without repositioning the patient. This adaptability improves the capacity to imagine physical designs according to alternate points of view.

4. **Utilitarian Imaging**

The consolidation of practical imaging strategies, for example, fMRI and dissemination weighted imaging, gives bits of knowledge into physiological and neurotic cycles. This useful data improves the demonstrative abilities of X-ray, particularly in neuroimaging and oncology.

5. **High-Goal Imaging**

Mechanical progressions have prompted the advancement of high-field X-ray frameworks, which proposition expanded signal-to-clamor proportions and worked on spatial goal. High-goal imaging is especially useful in applications, for example, outer muscle and neuroimaging, where definite representation is basic.

6. **Painlessness and Patient Solace**

X-ray is a painless imaging methodology that doesn't need the utilization of difference specialists by and large. The shortfall of ionizing radiation and the painless idea of the methodology add to patient solace and security. Be that as it may, contrast specialists might be utilized when upgraded perception is vital.

Difficulties and Constraints

1. **Cost and Openness**

X-ray frameworks are costly to gain, introduce, and keep up with. The expense of X-ray assessments can likewise be generally high. This has suggestions for openness, especially in asset restricted settings where the accessibility of cutting edge clinical imaging advancements might be restricted.

2. **Contraindications and Wellbeing Concerns**

Certain circumstances and clinical gadgets might be contraindications for X-ray. For instance, patients with pacemakers or certain metallic inserts may not be qualified for X-ray filters because of security concerns. Moreover, the solid attractive fields utilized in X-ray require severe wellbeing conventions to forestall mishaps connected with ferromagnetic items.

3. **Imaging Time**

X-ray assessments can be tedious, particularly for specific conventions that require various arrangements or specific imaging methods. Drawn out imaging times might present difficulties for patients who experience issues staying still or for those with claustrophobia.

4. **Challenges in Imaging Explicit Designs**

   Certain physical designs, like the lungs and bones, might be trying to picture with regular X-ray because of inborn limits. Procedures like attractive reverberation angiography (MRA) are utilized to envision veins, however imaging little designs with high spatial goal stays a test.

5. **Translation Difficulties**

The understanding of X-ray pictures requires specific preparation and mastery. Radiologists should be know about the subtleties of X-ray appearances and have the option to recognize ordinary varieties from obsessive circumstances. The intricacy of picture translation might affect the precision of analyses.

Future Headings and Developments

1. **Super High-Field X-ray**

   Headways in magnet innovation are driving the improvement of super high-field X-ray frameworks with field qualities surpassing 3 Tesla. These frameworks offer improved signal-to-commotion proportions and spatial goal, opening additional opportunities for nitty gritty imaging of physical designs and utilitarian cycles.

2. **Man-made reasoning (simulated intelligence) Incorporation**

   The reconciliation of computerized reasoning (man-made intelligence) and AI calculations into X-ray investigation is a quickly developing field. Artificial intelligence applications can aid picture understanding, computerize routine assignments, and work on the effectiveness and precision of finding. This incorporates mechanized injury recognition, picture division, and forecast of treatment reactions.

3. **Spectroscopy and Metabolic Imaging**

   Attractive reverberation spectroscopy (MRS) considers the evaluation of tissue digestion by breaking down the synthetic arrangement of tissues. Metabolic imaging with MRS has applications in oncology, nervous system science, and outer muscle imaging, giving bits of knowledge into cell processes and recognizing irregularities at a sub-atomic level.

4. **Atomic Imaging with X-ray**

   Progresses interestingly, specialists and imaging procedures are making ready for atomic imaging with X-ray. This includes picturing explicit particles or natural cycles inside the body, offering expected applications in malignant growth imaging, aggravation recognition, and checking restorative reactions at the sub-atomic level.

5. Utilitarian Connectomics

Utilitarian connectomics includes planning the associations and communications between various cerebrum locales utilizing progressed practical X-ray methods. This field holds guarantee for acquiring a more profound comprehension of mind capability, network modifications in neurological issues, and creating designated mediations.

Chapter 6

Challenges And Future Frontiers

In the steadily advancing scene of innovation, science, and society, difficulties and future outskirts arise as necessary parts that shape the direction of progress. From resolving squeezing worldwide issues to investigating strange domains, the transaction among difficulties and open doors moves mankind forward. This exhaustive investigation digs into the multi-layered domain of difficulties and future boondocks, analyzing the intricacies, distinguishing key areas of concern, and imagining the potential ways that lie ahead.

1. **Grasping Difficulties**
1. **Worldwide Wellbeing Difficulties**
1. Pandemic Readiness and Reaction
   The Coronavirus pandemic highlighted the basic requirement for worldwide pandemic readiness and reaction instruments. The difficulties presented by the quick spread of irresistible sicknesses featured the need of hearty medical services foundations, worldwide coordinated effort, and sped up antibody advancement processes. Tending to these difficulties requires a reconsideration of medical services frameworks, upgraded observation capacities, and proactive measures to moderate future pandemics.
2. **Wellbeing Imbalances**

Variations in medical services access and results endure around the world. Financial, geological, and segment factors add to wellbeing imbalances, affecting weak populaces excessively. Tending to these variations includes carrying out comprehensive medical services strategies, supporting medical care foundation in underserved areas, and advancing impartial dispersion of medical services assets.
2. **Ecological Difficulties**

1. **Environmental Change Relief and Variation**

   Environmental change stays perhaps of the main test confronting the planet. Relieving ozone depleting substance outflows, progressing to manageable energy sources, and carrying out versatile measures to address the effects of environmental change are basic goals. The convergence of science, strategy, and cultural activity is fundamental in diagramming a course toward a manageable and tough future.

2. **Biodiversity Misfortune**

The continuous loss of biodiversity presents dangers to environments, food security, and human prosperity. Deforestation, environment obliteration, and environmental change add to species elimination. Preservation endeavors, manageable land-use practices, and global participation are basic to save biodiversity and keep up with environmental equilibrium.

3. Mechanical Difficulties

1. **Online protection Dangers**

   As innovation turns out to be progressively interwoven with day to day existence, the gamble of online protection dangers raises. Cyberattacks on basic foundation, information breaks, and the abuse of arising innovations present huge difficulties. Fortifying online protection measures, encouraging worldwide collaboration, and creating versatile computerized frameworks are essential parts of tending to this advancing danger scene.

2. **Moral Issues in Computerized reasoning**

The quick headway of computerized reasoning (simulated intelligence) presents moral problems and cultural difficulties. Issues connected with predisposition in calculations, protection concerns, and the moral utilization of simulated intelligence in dynamic cycles require the improvement of powerful moral structures, straightforward practices, and mindful computer based intelligence administration.

4. Social and Financial Difficulties

1. **Disparity and Civil rights**

   Diligent social imbalances, including racial, orientation, and monetary variations, request thorough cultural changes. Advancing civil rights, cultivating inclusivity, and destroying foundational boundaries are urgent moves toward making an additional impartial and simply world.

2. **Monetary Interruptions**

Worldwide monetary vulnerabilities, exacerbated by occasions like the Coronavirus pandemic, highlight the requirement for strong financial frameworks. Building financial versatility includes enhancement, comprehensive arrangements, and inventive ways to deal with address disturbances and advance reasonable monetary development.

5. Instructive Difficulties

1. **Admittance to Quality Instruction**

   Differences in admittance to quality schooling endure around the world, impeding financial portability. Spanning the instructive hole requires interest in instructive foundation, utilizing innovation for remote learning, and executing comprehensive approaches that guarantee admittance to training for all.

2. **Adjusting to Mechanical Changes in Schooling**

The fast reconciliation of innovation in schooling presents difficulties connected with computerized proficiency, impartial admittance to innovation, and the advancing job of teachers. Adjusting school systems to use innovation successfully while tending to related difficulties is fundamental for getting ready people in the future for a mechanically determined world.

II. Exploring Future Outskirts

1. **Logical and Innovative Progressions**
1. Biotechnology and Accuracy Medication
   Headways in biotechnology, including CRISPR quality altering and customized medication, hold monstrous potential for reforming medical care. Accuracy medication, custom fitted to a person's hereditary cosmetics, guarantees more powerful medicines and designated mediations. Moral contemplations, administrative systems, and evenhanded admittance to these advancements will shape their effect on medical services.

2. **Quantum Figuring and Data Handling**

Quantum figuring addresses a wilderness that could change data handling capacities. The capacity to perform complex estimations dramatically quicker than old style PCs opens new roads in cryptography, advancement issues, and logical recreations. In any case, difficulties, for example, keeping up with quantum rationality and creating versatile quantum frameworks should be defeated for quantum registering to understand its maximum capacity.

2. Space Investigation and Colonization

1. **Investigation of Mars and Then some**

   The investigation of Mars remains as a critical boondocks in space investigation. Space offices and confidential elements are effectively dealing with missions to concentrate on the Martian climate and possibly lay out human states. Conquering difficulties connected with long-length space travel, life emotionally supportive networks, and the mental effect of confinement will be significant for the outcome of these undertakings.

2. **Economical Space Investigation**

As humankind grows its presence past Earth, supportable practices in space investigation become foremost. Tending to space trash, creating eco-accommodating drive frameworks, and guaranteeing mindful asset usage are fundamental for saving the space climate and cultivating long haul manageability.

3. **Environmentally friendly power and Economical Advances**

1. **Progressions in Sustainable power**

   The progress to sustainable power sources, for example, sunlight based and wind power, is a critical boondocks in tending to environmental change. Proceeded with progressions in energy capacity advancements, network joining, and the improvement of novel environmentally friendly power arrangements are basic for accomplishing a maintainable and low-carbon energy future.

2. **Round Economy Practices**

The idea of a round economy, zeroed in on limiting waste and expanding asset effectiveness, addresses a future wilderness in economical turn of events. Moving towards round economy rehearses includes reexamining item configuration, squander the board, and utilization examples to make a regenerative and shut circle framework.

4. **Human Increase and Man-made consciousness**

1. **Increased Reality and Augmented Reality Mix**

   The mix of expanded reality (AR) and computer generated reality (VR) into different parts of life, including medical services, training, and diversion, addresses an extraordinary wilderness. From vivid opportunities for growth to virtual clinical reenactments, AR and VR can possibly reshape how we interface with data and our general surroundings.

2. **Moral Contemplations in Human Expansion**

Progressions in human expansion advancements, for example, mind PC connection points and wearable gadgets, raise moral contemplations connected with security, assent, and the potential for unseen side-effects. Laying out moral structures and guidelines to direct the dependable turn of events and organization of these innovations is critical.

5. Natural Preservation and Reclamation

1. Reforestation and Environment Reclamation

The reclamation of debased environments and worldwide reforestation endeavors are basic for biodiversity preservation and environmental change alleviation. Utilizing innovation, like remote detecting and artificial intelligence, to screen and guide rebuilding drives, and executing arrangements that boost practical land the board, are imperative for the outcome of such undertakings.

2. Sea Protection and Economical Practices

Protecting the wellbeing of seas is a squeezing wilderness. Tending to difficulties, for example, overfishing, plastic contamination, and sea fermentation requires global participation, inventive innovations for reasonable fishing rehearses, and a promise to marine protection.

III. The Transaction Among Difficulties and Future Outskirts

1. Interconnected Nature of Difficulties

Challenges rarely exist in segregation; they are frequently interconnected and commonly supporting. For instance, natural difficulties, for example, environmental change can fuel wellbeing challenges by impacting the spread of irresistible illnesses and adding to asset shortage. An all encompassing methodology that considers the interconnected idea of difficulties is fundamental for successful critical thinking.

2. Development as an Impetus for Arrangements

Despite challenges, development arises as a strong impetus for arrangements. Mechanical progressions, logical forward leaps, and novel methodologies driven by human innovativeness can possibly address and defeat complex difficulties. Embracing a culture of development and cultivating coordinated effort between different disciplines are essential parts of exploring the intricate scene of difficulties.

3. Moral Contemplations in Development

As we explore future wildernesses, moral contemplations should direct development. Whether in the domains of biotechnology, man-made brainpower, or space investigation, moral systems guarantee that mechanical headways line up with human qualities, save individual freedoms, and contribute emphatically to cultural prosperity. Proactive moral conversations and the coordination of moral contemplations into innovative work processes are basic.

4. **Worldwide Cooperation and Organizations**

Tending to difficulties and investigating future outskirts requires worldwide joint effort and associations. Complex issues, for example, environmental change, pandemics, and innovative administration rise above public lines. Worldwide collaboration encourages the trading of information, assets, and skill, making an aggregate work to handle difficulties that influence the whole worldwide local area.

5. **Cultural Commitment and Inclusivity**

Cultural commitment and inclusivity assume critical parts in exploring difficulties and acknowledging future outskirts. Comprehensive dynamic cycles, different portrayal in exploration and development, and dynamic commitment with networks impacted by mechanical headways are fundamental for guaranteeing that the advantages of progress are shared impartially.

6.1Current challenges in understanding and manipulating magnetism

Attraction, a central power of nature, has enraptured researchers and specialists for quite a long time. The capacity to comprehend and control attraction has prompted extraordinary innovations as well as prepared for key experiences into the idea of issue. In the contemporary period, specialists are diving into multifaceted parts of attraction with an objective to open new wildernesses in innovation and extend how we might interpret the actual world. Nonetheless, this excursion isn't without its difficulties. In this far reaching investigation, we will unwind the on-going difficulties in understanding and controlling attraction, addressing the basic standards, mechanical applications, and the potential forward leaps that lie ahead.

1. **Essentials of Attraction**

1. **Quantum Nature of Attraction**

Attraction at its center is a quantum mechanical peculiarity. Understanding attraction at the quantum level includes wrestling with the way of behaving of electrons, turn cooperations, and the perplexing dance of attractive minutes inside materials. As materials therapist to nanoscale aspects, quantum impacts become progressively articulated, presenting intricacies that challenge customary models of attraction.

1. **Quantum Twist Fluids**

One of the fascinating peculiarities in quantum attraction is the idea of twist fluids. In specific materials, quantum changes can keep attractive minutes from adjusting into conventional arranged designs, prompting a state known as a twist fluid. Understanding and controlling twist fluids could have significant ramifications for both crucial physical science and applications in quantum registering.

2. Emanant Peculiarities in Attractive Materials

Attractive materials frequently show emanant peculiarities, where unforeseen properties rise out of the collaborations of rudimentary constituents. Models incorporate attractive skyrmions, which are nanoscale whirlpools of attractive request, and topological covers, which show novel electrical properties at their surfaces. Unwinding the hidden components of these developing peculiarities represents a huge test.

1. Skyrmions and Their Elements

Attractive skyrmions, with their expected applications in information capacity and spintronics, present a test in grasping their solidness, control, and dynamic way of behaving. Scientists are investigating ways of making and control skyrmions at room temperature, an essential step for commonsense applications.

3. Job of Quantum Snare

Quantum snare, a peculiarity where the conditions of at least two particles become entwined, assumes a part in figuring out attraction. In emphatically corresponded electron frameworks, where the way of behaving of electrons is profoundly caught, unwinding the impact of quantum ensnarement on attractive properties stays a considerable test.

1. Quantum Snare in Twist Frameworks

In turn frameworks, the ensnarement of quantum states turns out to be especially applicable. The ensnared quantum conditions of individual twists can prompt novel aggregate way of behaving, impacting the attractive properties of materials. Unraveling and bridling the force of quantum snare with regards to attraction is a state of the art area of examination.

II. Mechanical Difficulties in Attraction

1. Attractive Data Stockpiling and Handling

Attraction assumes a urgent part in data capacity and handling advancements. The scaling down of attractive stockpiling gadgets, for example, hard circle drives, faces difficulties as we approach the constraints of old style attractive recording innovations. The mission for higher capacity densities and quicker information handling has prompted the investigation of elective methodologies.

1. **Past Moore's Regulation in Attractive Capacity**

   Moore's Regulation, which depicts the multiplying of semiconductors on coordinated circuits roughly at regular intervals, has directed the advancement of data innovation. In the domain of attractive stockpiling, analysts are investigating post-Moore's Regulation advancements, including heat-helped attractive recording (HAMR) and shingled attractive recording (SMR), to conquer the impediments of customary attractive recording.

2. **Spintronics and Quantum Figuring**

Spintronics, a field that bridles the inherent twist of electrons for data handling, holds guarantee for the improvement of energy-effective and quicker figuring innovations. Quantum registering, utilizing the standards of quantum mechanics, presents the two potential open doors and difficulties in changing calculation. In any case, accomplishing shortcoming open minded quantum bits (qubits) and versatile quantum processors stays an impressive errand.

2. Attractive Materials for Energy Applications

Attraction is significant in different energy-related applications, including electric engines, generators, and sustainable power advances. The advancement of cutting edge attractive materials with further developed energy effectiveness, high-temperature solidness, and diminished uncommon earth component reliance is a squeezing challenge.

1. **Super durable Magnets for Practical Energy**

   Super durable magnets assume an imperative part in electric vehicles, wind turbines, and other green advances. The test lies in tracking down options in contrast to uncommon earth components, for example, neodymium and dysprosium, which are fundamental parts of elite execution extremely durable magnets however face issues of store network supportability and international worries.

2. **Attractive Refrigeration for Energy-Effective Cooling**

Attractive refrigeration, in light of the magnetocaloric impact, presents an eco-accommodating option in contrast to customary fume pressure refrigeration. In any case, creating savvy attractive refrigeration frameworks with wide relevance stays a test.

III. Challenges in Attraction Exploration

1. **Grasping Attractive Cooperations in Complex Frameworks**

Attraction in consolidated matter frameworks frequently includes complicated cooperations between different levels of opportunity, for example, orbital, cross section, and charge degrees. Figuring out the exchange of these connections in complex materials, including firmly corresponded electron frameworks, is a huge test.

1. Corresponded Electron Frameworks

In materials where electron cooperations rule, for example, high-temperature superconductors, disentangling the mind boggling transaction between charge, twist, and cross section levels of opportunity is a fantastic test. The unpredictable attractive and electronic stages showed by these materials stay at the very front of examination.

2. Ultrafast Polarization Elements

The ultrafast elements of polarization, happening on femtosecond timescales, present a test in exploratory and hypothetical examinations. Procedures, for example, femtosecond laser spectroscopy are utilized to test and control ultrafast charge processes, offering bits of knowledge into the principal timescales of attractive peculiarities.

1. Femtosecond Attraction and Spintronics

Understanding and controlling ultrafast polarization elements have suggestions for the field of spintronics, where the control of twist flows is utilized for data handling. Bridling femtosecond attraction can prompt progressions in ultrafast spintronic gadgets and memory advancements.

3. Attractive Geography and Quantum Materials

The investigation of topological parts of attraction, like topological encasings and topological magnonics, is a prospering field. Distinguishing and portraying new topological periods of issue, particularly those that display novel attractive properties, present both hypothetical and trial challenges.

1. Topological Magnonics

Topological magnonics investigates the topological properties of magnons, which are aggregate excitations of attractive minutes. The acknowledgment of topological magnonic states and their true capacity for data handling applications address continuous difficulties in the field of attraction.

4. Attractive Reverberation Imaging (X-ray) Difficulties

While Attractive Reverberation Imaging (X-ray) has reformed clinical diagnostics, challenges continue upgrading imaging goal, awareness, and explicitness.

Conquering these difficulties is fundamental for propelling clinical imaging capacities and opening new bits of knowledge into the human body.

1. High-Field X-ray and Imaging Goal

Progressions in high-field X-ray frameworks present difficulties connected with expanding attractive field qualities while moderating related specialized and security issues. Further developing imaging goal without compromising patient wellbeing stays a point of convergence in high-field X-ray research.

IV. Future Boondocks in Attraction

1. Quantum Advancements and Quantum Materials

The reconciliation of quantum innovations, including quantum processing and quantum correspondence, depends on the improvement of new quantum materials. Quantum magnets, superconductors, and topological separators are at the very front of examination, holding the possibility to change data handling and correspondence.

1. Quantum Twist Fluids in Quantum Processing

Investigating the properties of quantum turn fluids might prepare for novel quantum processing standards. The fascinating quantum states showed by turn fluids could offer novel benefits in quantum data capacity and handling.

2. Attractive Nanomaterials and Nanotechnology

Headways in nanotechnology are opening new wildernesses in the control of attraction at the nanoscale. Attractive nanomaterials, for example, nanoparticles and nanocomposites, hold guarantee for applications going from designated drug conveyance to high-thickness information capacity.

1. Nanomagnetic Gadgets for Data Stockpiling

The improvement of nanomagnetic gadgets, including spintronic gadgets and attractive irregular access memory (MRAM), is a boondocks that means to beat the restrictions of regular semiconductor-based innovations. Outfitting the extraordinary properties of attractive nanoparticles may prompt forward leaps in energy-productive registering.

3. Quantum Materials for Energy Applications

The journey for quantum materials with fitted attractive and electronic properties stretches out to energy-related applications. Investigating new materials for proficient magnetocaloric cooling, thermoelectric change, and energy collecting addresses a boondocks in economical energy research.

1. Topological Materials for Energy-Proficient Hardware

Topological materials, described by one of a kind electronic band structures, hold potential for creating energy-productive electronic gadgets. Saddling the topological properties of materials could prompt progressions in low-power gadgets and quantum data handling.

4. Attractive Reverberation Advances Past Imaging

The utilizations of attractive reverberation reach out past conventional imaging modalities. Arising advances influence attractive reverberation for novel purposes, including quantum detecting, materials portrayal, and in any event, investigating the secrets of the universe.

1. Quantum Detecting and Attractive Reverberation

Quantum sensors in light of attractive reverberation standards are being produced for ultrasensitive estimations. These sensors have applications in fields going from clinical diagnostics to the location of tricky peculiarities in basic material science.

6.2 Cutting-edge research in magnetic materials

Quantum registering, a progressive worldview in data handling, has arisen as an outskirts that holds the commitment of changing different enterprises and logical disciplines. The exceptional standards of quantum mechanics, for example, superposition and entrapment, make ready for computational capacities that far outperform traditional PCs. Past quantum registering, these standards likewise track down applications in trend setting innovations that reach from secure correspondence to materials science. In this extensive investigation, we will dig into the expected uses of quantum registering and trend setting innovations, disentangling the extraordinary effect they could have on the manner in which we figure out, process, and use data.

1. Quantum Registering: A Change in outlook in Data Handling

1. Prologue to Quantum Registering

Quantum registering tackles the standards of quantum mechanics to perform calculations in manners essentially not quite the same as old style figuring. Traditional pieces, addressing 0s and 1s, are supplanted by quantum bits or qubits, which can exist in superposition conditions of both 0 and 1 at the same time. This remarkable component empowers quantum PCs to investigate different conceivable outcomes in equal, dramatically expanding their computational power for specific errands.

2. **Quantum Calculations and Speedup**

1. **Shor's Calculation for Whole number Factorization**

   One of the notable utilizations of quantum registering is Shor's calculation, which productively factors huge numbers dramatically quicker than the most popular old style calculations. The capacity to factor huge numbers rapidly has suggestions for breaking broadly utilized cryptographic conventions, like RSA, presenting both a danger to network protection and an inspiration for the improvement of quantum-safe cryptographic techniques.

2. **Grover's Calculation for Unstructured Inquiry**

Grover's calculation improves the proficiency of unstructured hunt issues. In traditional figuring, looking through an unsorted data set demands straight investment, yet Grover's calculation accomplishes a quadratic speedup. This calculation has suggestions for cryptographic applications, improvement issues, and information base looking.

3. Quantum Matchless quality and Then some

1. **Accomplishing Quantum Matchless quality**

   Quantum matchless quality alludes direct at which a quantum PC can beat the most exceptional traditional supercomputers in unambiguous errands. In 2019, Google professed to have accomplished quantum matchless quality with their 53-qubit Sycamore processor, exhibiting the capability of quantum PCs to take care of mind boggling issues.

2. **Difficulties and Adaptability**

While quantum matchless quality addresses a huge achievement, challenges stay in accomplishing functional quantum figuring applications. Blunder adjustment, sound decrease, and keeping up with quantum intelligence over longer lengths are basic variables for scaling quantum PCs to handle true issues.

II. Uses of Quantum Processing

1. **Cryptography and Security**

1. **Quantum Key Appropriation (QKD)**

   Quantum registering's effect on cryptography reaches out past danger vectors to expected arrangements. Quantum Key Conveyance (QKD) use the standards of quantum mechanics to empower secure correspondence channels, distinguishing any listening in endeavors. QKD offers a quantum-safe way to deal with secure correspondence.

2. **Post-Quantum Cryptography**

The coming of strong quantum calculations represents a danger to existing cryptographic frameworks. Present quantum cryptography points on foster cryptographic calculations that stay secure against quantum assaults. The examination and execution of post-quantum cryptographic principles are significant for guaranteeing the drawn out security of delicate data.

2. Advancement and Recreation

1. **Quantum AI**

   Quantum PCs can possibly alter AI calculations. Quantum AI investigates the utilization of quantum frameworks to improve advancement assignments, information examination, and example acknowledgment. Quantum calculations, for example, the Quantum Backing Vector Machine, exhibit the potential for dramatic speedup in AI errands.

2. **Quantum Recreations for Material Science**

Mimicking the way of behaving of particles and materials at the quantum level is a computationally requesting task. Quantum PCs can offer huge benefits in recreating sub-atomic designs, permitting scientists to investigate new materials, plan sedates, and figure out compound responses with exceptional exactness.

3. Money and Portfolio Advancement

1. **Quantum Money Models**

Quantum PCs show guarantee in monetary applications, especially in portfolio advancement and hazard the board. Quantum calculations can effectively break down complex monetary informational collections, advancing venture portfolios and relieving monetary dangers. Quantum figuring's capacity to deal with huge datasets at the same time gives a critical benefit in monetary demonstrating.

4. Inventory network and Operations

1. **Quantum Improvement for Operations**

Quantum calculations, for example, the Quantum Rough Enhancement Calculation (QAOA), can address complex streamlining issues intrinsic in store network and strategies. From course enhancement to stock administration, quantum processing offers the potential for additional proficient and financially savvy arrangements.

5. Medical services and Medication Revelation

1. **Quantum Science Reproductions**

   Quantum PCs are ready to reform drug disclosure by reenacting atomic connections with extraordinary precision. Quantum science reproductions can display complex natural frameworks, speeding up the medication disclosure process and possibly prompting the improvement of new medicines for different infections.

2. **Improving Clinical Preliminaries**

Quantum registering's enhancement capacities can be applied to smooth out the plan and execution of clinical preliminaries. By dissecting immense datasets and advancing preliminary boundaries, quantum calculations can add to more proficient and savvy clinical exploration.

6. Man-made brainpower and Brain Organizations

1. **Quantum Brain Organizations**

Quantum PCs can possibly improve man-made consciousness (simulated intelligence) calculations through the advancement of quantum brain organizations. Quantum brain organizations can process and examine complex informational collections all the more effectively, offering benefits in errands like example acknowledgment and picture characterization.

7. Natural Displaying and Environmental Change Studies

1. **Quantum Figuring for Environment Demonstrating**

Tending to the intricacies of environment displaying and understanding the effect of human exercises on the climate requires huge computational power. Quantum PCs can add to more exact and definite environment recreations, helping researchers in creating procedures to moderate the impacts of environmental change.

8. Traffic Improvement and Metropolitan Preparation

1. **Quantum Calculations for Traffic Stream**

Quantum registering can be applied to enhance traffic stream in metropolitan regions. By breaking down traffic designs, quantum calculations can recommend ideal courses and timing for traffic lights, adding to diminished blockage and further developed transportation productivity.

9. Game Hypothesis and Independent direction

1. **Quantum Game Hypothesis**

Quantum registering acquaints novel methodologies with game hypothesis and direction. Quantum game hypothesis investigates techniques that influence quantum ensnarement and superposition to accomplish results that are unrealistic in traditional game hypothesis, opening new roads for key preparation.

III. Past Quantum Registering: Cutting edge innovations

1. **Quantum Correspondence**
1. **Quantum Key Appropriation (QKD) Organizations**

   Quantum correspondence stretches out past cryptographic applications. Quantum Key Appropriation (QKD) networks empower secure correspondence channels between various gatherings, utilizing the standards of quantum mechanics to trade cryptographic keys with remarkable security.
2. **Quantum Web**

The vision of a quantum web includes an organization where quantum data can be communicated safely over significant distances. Quantum ensnarement empowers the production of quantum snare based networks, cultivating secure correspondence and appropriated quantum processing.

2. Quantum Detecting and Metrology

1. **Quantum Sensors for Accuracy Estimations**

   Quantum sensors influence the standards of quantum mechanics to accomplish phenomenal degrees of accuracy in estimations. Quantum metrology applications incorporate super exact timekeeping, attractive field detecting, and gravitational field estimations.
2. **Quantum Imaging**

Quantum imaging strategies, for example, quantum-upgraded imaging and quantum radar, exploit quantum snare and superposition to accomplish higher goal and responsiveness contrasted with traditional imaging techniques. Quantum imaging has applications in clinical diagnostics, observation, and remote detecting.

3. Quantum Materials and Nanotechnology

1. **Quantum Materials for Hardware**

   Quantum materials, including topological separators and superconductors, have applications in cutting edge gadgets. These materials display novel electronic properties that can be outfit for creating energy-productive electronic gadgets and quantum PCs.
2. **Nanotechnology and Quantum Specks**

Quantum specks, nanoscale semiconductor particles, display quantum mechanical properties. These nanomaterials have applications in shows, sun oriented cells, and clinical imaging, where their quantum conduct considers exact command over optical and electronic properties.

4. Quantum Biophysics and Medication

1. Quantum Science

Quantum science investigates the job of quantum peculiarities in organic frameworks. Quantum impacts in cycles, for example, photosynthesis and magnetoreception could have suggestions for the advancement of novel clinical treatments and biotechnological applications.

2. Quantum-upgraded Clinical Imaging

Quantum-upgraded clinical imaging strategies, including quantum-upgraded attractive reverberation imaging (QMRI), plan to further develop imaging goal and awareness. Quantum sensors and finders could upgrade the abilities of clinical symptomatic instruments, prompting more exact and productive medical care rehearses.

5. Quantum Advanced mechanics and Man-made brainpower

1. Quantum AI for Advanced mechanics

Quantum AI calculations have applications in mechanical technology, empowering machines to handle complex informational collections and go with choices all the more proficiently. Quantum-upgraded man-made reasoning can add to headways in independent frameworks and automated advances.

2. Quantum Mechanical technology for Sensor Organizations

Quantum mechanical technology investigates the reconciliation of quantum sensors and correspondence in automated frameworks. This interdisciplinary field plans to foster automated networks with upgraded detecting abilities, empowering applications in natural checking, debacle reaction, and observation.

6. Quantum-improved Energy Innovations

1. Quantum-improved Sun powered Cells

Quantum-upgraded approaches in materials science can add to the advancement of additional productive sun based cells. By utilizing quantum standards, scientists expect to improve light retention and energy change effectiveness in photovoltaic gadgets.

2. Quantum Thermoelectric Gadgets

Quantum thermoelectric gadgets, which convert temperature contrasts into electrical energy, have possible applications in energy gathering and waste intensity recuperation. Quantum impacts in materials could prompt the advancement of additional productive thermoelectric generators.

IV. Difficulties and Contemplations

1. Specialized Difficulties in Quantum Registering

1. Blunder Remedy and Quantum Decoherence

Keeping up with the solidness of qubits is a huge test in quantum registering. Quantum decoherence, the deficiency of quantum data because of ecological communications, represents a danger to the dependability of quantum calculations. Blunder remedy procedures are being created to address these difficulties.

2. Quantum Equipment Improvement

Headways in quantum registering depend on the improvement of strong quantum equipment. Building versatile and shortcoming open minded quantum processors is a complicated errand that requires defeating specialized difficulties in material science, creation, and control of quantum frameworks.

2. Security and Moral Contemplations

1. Quantum-safe Cryptography

The turn of events and arrangement of quantum-safe cryptographic norms are fundamental to shield delicate data from potential quantum assaults. Changing to quantum-safe encryption strategies is a basic move toward guaranteeing the drawn out security of computerized correspondence.

2. Moral Contemplations in Quantum Advances

As quantum innovations advance, moral contemplations with respect to protection, security, and the likely abuse of quantum figuring capacities become fundamental. Laying out moral rules and capable practices in the turn of events and arrangement of quantum advancements is fundamental.

3. Incorporation with Traditional Frameworks

1. Half and half Quantum-Traditional Frameworks

Coordinating quantum advances with existing traditional frameworks presents difficulties regarding similarity and correspondence. Creating crossover quantum-old style frameworks that really influence the qualities of the two ideal models is urgent for the reasonable execution of quantum advances.

2. Quantum Correspondence Framework

Building a quantum correspondence framework for huge scope quantum networks requires the improvement of solid quantum repeaters, quantum recollections, and different parts. Laying out a worldwide quantum correspondence network faces specialized, calculated, and normalization challenges.

V. Future Boondocks and Conceivable outcomes

1. Quantum Processing Past von Neumann Engineering

1. Quantum Neuromorphic Processing

Investigating neuromorphic processing models that join quantum and old style components could prompt more proficient calculations for computerized reasoning and AI. Quantum neuromorphic registering expects to mirror the mind's equal handling abilities for further developed design acknowledgment and navigation.

2. Quantum Web and Entrapment based Organizations

1. Entrapment based Quantum Organizations

Progressing towards the acknowledgment of a quantum web includes the improvement of ensnarement based quantum organizations. These organizations could empower secure correspondence, circulated quantum processing, and novel applications in quantum-improved advances.

3. Quantum-improved Detecting Advances

1. Quantum-improved Imaging and Detecting Gadgets

Proceeded with research in quantum detecting advances means to foster more delicate and exact gadgets. Quantum-upgraded sensors could track down applications in clinical diagnostics, natural observing, and different fields where high-accuracy estimations are urgent.

4. Quantum-improved Materials Revelation

1. Quantum Processing for Materials Science

Quantum processing is ready to alter materials science by speeding up the disclosure and plan of new materials. Quantum calculations can investigate tremendous material spaces, anticipate properties, and guide trial endeavors, prompting the improvement of cutting edge materials with custom-made functionalities.

5. Quantum-upgraded Energy Stockpiling

1. Quantum-upgraded Batteries and Energy Stockpiling

Investigating quantum-upgraded approaches in energy capacity includes utilizing quantum standards for additional effective batteries. Quantum advancements might add to the improvement of cutting edge energy capacity gadgets with further developed limit, life expectancy, and charging speeds.

6. Quantum-improved artificial intelligence and Advanced mechanics

1. Quantum-improved AI for Mechanical technology

Quantum-improved AI calculations could track down applications in mechanical technology, empowering more versatile and shrewd automated frameworks. Coordinating quantum abilities with mechanical innovations might prompt progressions in independence, navigation, and sensor organizations.

7. Quantum-upgraded Natural Observing

1. Quantum Sensors for Ecological Investigations

Quantum-upgraded sensors can possibly change natural observing. From identifying poisons to concentrating on biological systems, quantum sensors could give phenomenal experiences into the intricacies of the regular world.

8. Quantum-upgraded Medical care Advancements

1. Quantum-upgraded Clinical Imaging Gadgets

Proceeded with headways in quantum-improved clinical imaging could reform diagnostics and treatment arranging in medical services. Quantum sensors and imaging advancements might empower higher goal, responsiveness, and explicitness in clinical imaging applications.

9. Quantum-upgraded Materials for Hardware

1. Quantum Gadgets and Spintronics

Investigating quantum-upgraded materials for hardware includes bridling quantum properties for growing more productive electronic gadgets. Quantum spintronics, specifically, expects to use quantum turn for data handling and stockpiling, prompting energy-effective and quicker figuring advances.

Chapter 7

Iron In Everyday Life

Iron, a flexible and bountiful metal, assumes a significant part in forming the texture of our daily existences. From the development of famous designs to the working of fundamental organic cycles, iron is a principal component that spans the domains of industry, innovation, and science. This thorough investigation digs into the complex presence of iron in our everyday presence, disentangling its commitments to framework, innovation, wellbeing, and then some.

1. **Prologue to Press**
1. **Synthetic and Actual Properties of Iron**

   Iron, with the image Fe and nuclear number 26, has a place with the progress metal gathering on the occasional table. It is known for its trademark shine, flexibility, and attractive properties. In different oxidation states, iron structures a huge number of mixtures, displaying its flexibility in compound cooperations.
2. **Overflow and Extraction of Iron**

Iron is quite possibly of the most plentiful component on The planet, comprising a huge part of the World's center. The essential wellsprings of iron incorporate iron minerals, for example, hematite and magnetite, which are extricated through mining and refined through different metallurgical cycles, dominatingly the impact heater strategy.

II. Iron in Development and Framework

1. **Underlying Steel and Structures**

1. **Job of Iron in Development**

   Iron, as underlying steel, is a foundation of present day development. Its high strength and toughness make it an optimal material for developing structures, spans, and different foundation projects. The utilization of iron in development goes back hundreds of years and keeps on being a characterizing component in structural and designing undertakings.

2. **Famous Designs Made of Iron**

A few famous designs all over the planet grandstand the unmistakable job of iron in development. The Eiffel Pinnacle in Paris, the Door Curve in St. Louis, and the Iron Scaffold in Shropshire, Britain, stand as demonstrations of the getting through strength and tasteful allure of iron-based structures.

2. Transportation and Iron Compounds

1. **Iron in Car Industry**

   The car business vigorously depends on endlessly iron combinations, especially steel. The undercarriage, motor parts, and different primary components of autos are produced using iron-based materials. The toughness and cost-viability of iron go with it a favored decision in vehicle producing.

2. **Rail lines and Iron Tracks**

Rail transportation, an imperative part of worldwide strategies, depends on iron for the development of rail line tracks. The toughness and high elasticity of iron compounds guarantee the life span and security of railroad framework, working with proficient and dependable transportation.

3. Iron in Extension Development

1. **Job of Iron in Extension Designing**

   Iron assumes a vital part in the development of extensions, giving the important solidarity to help weighty loads and endure ecological variables. Engineered overpasses, curve extensions, and pillar spans frequently integrate iron and steel parts, guaranteeing primary trustworthiness and life span.

2. **Instances of Iron Scaffolds**

Unmistakable iron extensions, for example, the Brooklyn Scaffold in New York and the Forward Extension in Scotland, outline the designing ability and solidness of iron in span development. These designs stand as designing wonders that have endured everyday hardship.

III. Iron in Innovation and Development

1. **Attractive Properties of Iron**
1. **Ferromagnetism and Attractive Stockpiling**

   Iron is ferromagnetic, meaning it displays major areas of strength for a property when presented to an attractive field. This trademark is saddled in different mechanical applications, especially in attractive capacity gadgets. Attractive tapes, hard drives, and other stockpiling media depend on iron-based materials to store and recover information.
2. **Job of Iron in Electromagnetism**

The attractive properties of iron add to the usefulness of electromagnets. By passing an electric flow through a curl containing iron, an electromagnet can be made. This guideline is applied in different gadgets, including electric engines, speakers, and attractive reverberation imaging (X-ray) machines.

2. Gadgets and Iron-based Parts

1. **Iron in Electrical Wiring**

   Iron and its combinations, like steel, are essential parts in the development of electrical wiring. The conductivity and pliability of iron-based materials make them reasonable for communicating electrical flows, supporting the tremendous organization of force conveyance frameworks.
2. **Transformers and Inductors**

Transformers and inductors, fundamental parts in electrical frameworks, frequently contain iron centers. The attractive properties of iron work with the change of electrical energy, taking into consideration effective power conveyance and voltage guideline.

3. Iron in Broadcast communications

1. **Job in Media transmission Links**

The transmission of broadcast communications signals depends on iron-based materials for the development of links. Iron centers offer underlying help and add to flag respectability, guaranteeing dependable correspondence over significant distances.

4. Iron in Processing and Server farms

1. **Information Capacity Gadgets**

   Iron's attractive properties assume a significant part in the improvement of information stockpiling gadgets. From early attractive tapes to current hard drives, iron-based materials add to the effective and minimized capacity of huge measures of advanced data.

2. **Server Foundation**

Server farms, the foundation of computerized administrations and distributed computing, frequently consolidate endlessly iron amalgams in server racks and underlying parts. The strength and solidness of iron add to the dependability and unwavering quality of these basic figuring conditions.

IV. Iron in Natural Frameworks

1. Job of Iron in Human Science

1. Iron in Hemoglobin and Oxygen Transport

   Iron is a fundamental component for human wellbeing, assuming a focal part in the design of hemoglobin. Hemoglobin, present in red platelets, ties to oxygen in the lungs and transports it all through the body, working with cell breath.

2. **Iron in Myoglobin and Muscle Capability**

Myoglobin, a protein like hemoglobin, contains iron and is liable for putting away and shipping oxygen in muscle tissues. Iron's contribution in myoglobin guarantees sufficient oxygen supply to muscles during actual work.

2. Lack of iron and Paleness

1. **Pervasiveness and Reasons for lack of iron**

   Lack of iron is a typical healthful concern around the world, prompting conditions like pallor. Deficient admission of iron-rich food varieties, unfortunate ingestion, or expanded iron necessities can add to press lack.

2. **Iron Supplementation and Treatment**

Iron supplementation and dietary changes are normal methodologies for treating lack of iron and paleness. Iron enhancements, frequently as ferrous sulfate, mean to reestablish iron levels in the body and mitigate side effects related with low iron.

3. Iron in Plant Science

1. **Job in Photosynthesis and Chlorophyll**

   While iron is definitely not a significant underlying part in plant cells, it assumes a pivotal part in photosynthesis. Iron is engaged with the union of chlorophyll, the green color answerable for catching light energy during photosynthesis.

2. **Iron Take-up and Supplement Transport in Plants**

Plants have multifaceted systems for iron take-up from the dirt. Iron is fundamental for different cell processes in plants, including supplement transport and chemical enactment.

V. Social and Authentic Meaning of Iron

1. **Iron Age and Human Civilization**
1. **Progress to the Iron Age**

   The revelation and dominance of iron metallurgy denoted a huge achievement in mankind's set of experiences, prompting the Iron Age. This time saw the far and wide utilization of iron instruments, weapons, and executes, changing horticulture, development, and fighting.
2. **Archeological Antiquities and Iron Masterfulness**

Archeological unearthings have uncovered iron antiquities that give experiences into old societies and their innovative progressions. Complicatedly created iron devices, weapons, and craftsmanship objects feature the imaginativeness and expertise of old ironworkers.

2. Imagery and Folklore

1. **Emblematic Meaning of Iron**

   Iron holds emblematic importance in different societies and legends. Frequently connected with strength, solidness, and craftsmanship, iron is an image of versatility and guts.
2. **Iron in Old stories and Legends**

Various folktales and legends highlight iron as a focal component, underlining its extraordinary and defensive characteristics. From iron weapons that kill legendary monsters to press horseshoes accepted to bring best of luck, these accounts mirror the social importance ascribed to press.

VI. Difficulties and Contemplations

1. **Ecological Effect of Iron Extraction**
1. **Mining and Asset Exhaustion**

The extraction of iron mineral through mining processes raises worries about natural debasement and asset consumption. Feasible mining practices and endeavors to reuse and reuse iron-based materials intend to alleviate these ecological effects.

2. Consumption and Material Corruption

1. **Rust Arrangement in Iron-based Designs**

Iron and steel are helpless to consumption, with rust development being a typical issue. Erosion compromises the primary honesty of iron-based structures, requiring defensive measures like coatings and galvanization.

3. Wellbeing Difficulties Connected with Iron Over-burden

1. Hemochromatosis and Iron Amassing

While lack of iron is a pervasive wellbeing concern, extreme iron admission or conditions, for example, hemochromatosis can prompt iron over-burden. This can bring about organ harm and different unexpected issues, featuring the significance of adjusted iron levels.

4. Innovative Advances and Material Other options

1. Investigation of Elective Materials

Headways in materials science and designing have prompted the investigation of elective materials with properties tantamount to or outperforming those of iron. From composite materials to cutting edge amalgams, scientists look for choices that deal further developed strength, decreased weight, and improved maintainability.

VII. Future Viewpoints and Developments

1. Shrewd Materials and Nanotechnology

1. Job of Iron in Shrewd Materials

Headways in shrewd materials and nanotechnology include the mix of iron nanoparticles for creative applications. These applications incorporate designated drug conveyance, detecting advancements, and the improvement of versatile materials with tunable properties.

2. Biomedical Utilizations of Iron Nanoparticles

1. Iron Nanoparticles in Medication

Iron nanoparticles, especially as iron oxide, certainly stand out enough to be noticed for their biomedical applications. These incorporate designated drug conveyance, attractive reverberation imaging (X-ray) contrast specialists, and hyperthermia treatment for malignant growth treatment.

3. Joining of Iron in Supportable Advancements

1. Environmentally friendly power and Iron-based Parts

Iron's job stretches out into the domain of sustainable power advancements. From iron-containing impetuses for hydrogen creation to press based materials in sun powered cells, scientists investigate feasible applications that tackle the properties of iron for clean energy arrangements.

4. Progressions in Iron Metallurgy

1. Developments in Iron Creation Cycles

Progressing research in metallurgy means to upgrade iron creation processes, making them more energy-proficient and ecologically maintainable. Developments in refining advances and elective decrease strategies add to the advancement of iron metallurgy.

7.1Iron's role in everyday objects and appliances

Iron, a universal and flexible metal, pervades our regular routines in manners frequently inconspicuous. From the primary parts of structures to the internal operations of home devices, iron's presence is basic to the usefulness and sturdiness of a bunch of ordinary items. This thorough investigation digs into the assorted utilizations of iron in different family things and machines, disentangling the critical job it plays in molding the advanced accommodations we frequently underestimate.

1. Presentation

1. Meaning of Iron in Day to day existence

Iron, with its overflow, strength, and different properties, holds a vital spot in the texture of our day to day presence. Its applications in ordinary items and apparatuses range various businesses, from development to assembling, and add to the effectiveness and unwavering quality of items we use consistently.

2. Primary and Practical Qualities of Iron

Iron's substance and actual properties, including its pliability, conductivity, and attractive nature, make it an optimal material for a large number of utilizations. As we explore the different areas of regular day to day existence, we experience iron in structures that fill both underlying and utilitarian needs.

II. Iron in Development and Home Foundation

1. Iron in Building Designs

1. Primary Steel in Development

One of the essential utilizations of iron in regular daily existence is in the development of structures. Primary steel, overwhelmingly made out of iron, gives the structure to high rises, private structures, and different framework projects. Its high strength and sturdiness add to the solidness and life span of designs.

2. **Support in Concrete**

Iron rebar, short for supporting bar, is broadly used to build up substantial designs. Its incorporation upgrades the rigidity of cement, guaranteeing the primary honesty of extensions, parkways, and structures. The mix of iron and substantial structures a considerable pair in development.

2. Iron in Home Machines

1. **White Merchandise: Clothes washers and Coolers**

   Home machines, regularly alluded to as white products, frequently contain iron parts. Clothes washers, for instance, consolidate iron as a hearty drum that endures the afflictions of rehashed use. Coolers use iron in the development of racks, drawers, and the blower, adding to their usefulness and life span.

2. **Iron in Kitchen Machines**

The kitchen is a center point of iron-containing machines. From cookware like cast iron skillets to little machines, for example, toaster ovens and blenders, iron is a fundamental part. The strength and intensity maintenance properties of iron make it a favored material for different culinary devices.

3. Iron in Furnishings and Home Style

1. **Fashioned Iron Furnishings**

   Fashioned iron, a type of iron with an exceptionally low carbon content, tracks down its place in furnishings and home stylistic layout. Created iron furnishings, known for its rich plans and solidness, graces outside spaces, nurseries, and insides, adding both usefulness and stylish enticement for homes.

2. **Iron Components in Home Stylistic theme**

Iron's adaptability stretches out to beautiful components in homes, including railings, light apparatuses, and decorative pieces. The utilization of iron in home stylistic layout consolidates structure and capability, offering both primary help and visual interest.

4. Iron in Home Improvement Devices

1. **Hand Devices and Power Instruments**

   Home improvement projects depend vigorously on iron-based devices. Hammers, screwdrivers, wrenches, and drive instruments frequently have iron parts, utilizing its solidarity and sturdiness to deal with different assignments. Iron's presence in these devices adds to their life span and adequacy.

2. Iron Latches: Nails and Screws

Nails and screws, fundamental parts in carpentry and development, are normally produced using iron or iron combinations. Their job in consolidating materials features the crucial significance of iron in the development and gathering of ordinary items.

III. Iron in Transportation and Portability

1. Autos and Iron Compounds

1. Primary Parts in Vehicles

The auto business broadly uses endlessly iron combinations in the assembling of vehicles. The undercarriage, motor blocks, and different underlying parts benefit from iron's high strength and effect opposition, adding to the security and dependability of cars.

2. Iron in Auto Motors

The gas powered motors that power most vehicles depend on iron for parts, for example, chamber blocks and driving rods. Iron's intensity opposition and mechanical properties make it a reasonable material for the requesting conditions inside a motor.

2. Iron in Bikes and Bikes

1. Outline Development in Bikes

Bikes, a well known method of transportation, frequently have outlines produced using iron or iron compounds. The power and reasonableness of iron add to the far and wide utilization of this material in bike development, guaranteeing the solidness of the casing under different riding conditions.

2. Cruisers and Iron Parts

Cruisers, going from worker bicycles to superior execution models, consolidate iron in different parts. The edge, motor housings, and other underlying components benefit from iron's solidarity, making cruisers dependable and strong.

3. Rail Transportation and Iron Tracks

1. Rail route Tracks and Framework

Rail transportation, a basic part of worldwide portability, depends on iron for the development of rail line tracks. Iron's sturdiness and high rigidity guarantee the life span of rail route framework, supporting the productive and safe development of trains.

2. Iron in Trains and Train Parts

Trains and prepare vehicles contain various iron parts. From the steel wheels that roll on iron tracks to the motor parts, iron is essential to the development of trains, adding to their proficiency and primary honesty.

4. Aviation Utilizations of Iron

1. Primary Parts in Airplane

The airplane business uses iron combinations, especially tempered steel, in the development of airplane. Primary parts, landing stuff, and motor parts benefit from the solidarity to-weight proportion and erosion opposition of iron-based materials.

IV. Iron in Innovation and Gadgets

1. Attractive Properties of Iron in Gadgets

1. Attractive Capacity Gadgets

Iron's attractive properties assume a urgent part in the field of gadgets. Attractive capacity gadgets, including hard drives and attractive tapes, use iron-based materials to store and recover advanced data. The attractive polarization of iron particles empowers the encoding of information in these capacity mediums.

2. Transformers and Inductors

Iron's attractive qualities add to the usefulness of transformers and inductors in electronic circuits. These parts, urgent for voltage guideline and sign handling, frequently contain iron centers that work with the proficient exchange of electromagnetic energy.

2. Iron in Buyer Hardware

1. Cell phones and Iron Parts

Buyer hardware, for example, cell phones, consolidate iron in different parts. The underlying edge, connectors, and inside parts benefit from the utilization of iron composites, giving both toughness and electromagnetic protecting.

2. PCs and Iron in Handling Units

PCs and servers, the foundation of present day figuring, contain iron-based parts. From the skeleton to the processor, iron adds to the underlying uprightness and warm administration of these electronic gadgets.

3. Iron in Correspondence Advancements

1. **Fiber Optic Organizations and Iron Foundation**
   Correspondence organizations, including the huge fiber optic foundation that empowers rapid web, frequently depend on iron parts for help designs and nooks. Iron's strength guarantees the unwavering quality of these basic correspondence frameworks.
2. **Satellites and Space Correspondence**

Satellites, fundamental for worldwide correspondence and route, consolidate iron composites in their underlying parts. The cruel states of room request materials with a high solidarity to-weight proportion, settling on iron an invaluable decision for satellite development.

4. Sustainable power Advancements and Iron

1. **Iron in Wind Turbines**
   Wind turbines, a critical part of sustainable power frameworks, use iron in different parts. The underlying scaffolding, cog wheels, and generator housings benefit from the strength and solidness of iron composites, adding to the proficiency of wind energy age.
2. **Sunlight powered chargers and Iron Mounting Designs**

Sun powered chargers, collecting energy from daylight, frequently depend on iron for mounting structures. The help approaches that position sun powered chargers to catch daylight at ideal points are every now and again developed from iron, guaranteeing solidness and life span.

V. Iron in Home devices

1. **Iron in Kitchen Apparatuses**
1. **Cookware: Pots and Container**
   Cookware, a fundamental piece of each and every kitchen, regularly incorporates iron-based things. Project iron pots and container, celebrated for their in any event, warming and intensity maintenance properties, have been a staple in kitchens for ages.
2. **Little Kitchen Apparatuses**

Little kitchen apparatuses, from toaster ovens to blenders, frequently contain iron parts. The primary components and lodging of these apparatuses benefit from the strength and toughness of iron, guaranteeing their usefulness and life span.

2. **Clothing Apparatuses: Irons and Clothes washers**

1. **Garments Irons and Pressing Sheets**

   The exemplary garments iron, a family staple for piece of clothing care, is commonly developed with iron components. Pressing sheets, intended to give a steady surface to pressing, likewise consolidate iron in their edges.

2. **Clothes washers and Iron Parts**

Clothes washers, imperative for present day clothing schedules, contain iron parts in their primary edges and inward components. Iron's solidness guarantees that these apparatuses endure the mechanical burdens related with rehashed use.

3. **Warming and Cooling Machines**

1. **Radiators and Warming Frameworks**

   In warming frameworks, radiators frequently contain iron or iron amalgams because of their phenomenal intensity conduction properties. Iron's capacity to endure high temperatures makes it appropriate for the productive exchange of intensity in home warming frameworks.

2. **Forced air systems and Iron Curls**

Cooling units use iron curls in their intensity trade frameworks. The warm conductivity of iron backings the compelling exchange of intensity, adding to the cooling productivity of these apparatuses.

4. **Electronic Contraptions and Iron Parts**

1. **TVs and Iron in Packaging**

   TVs and screens frequently highlight iron parts in their housings. Iron's underlying properties offer the important help and sturdiness for these electronic devices.

2. **Home Theater setups and Iron in Speakers**

Speakers, vital to home theater setups, consolidate iron parts in their attractive gatherings. The collaboration among iron and magnets empowers the transformation of electrical signs into sound waves.

VI. **Iron in Wellbeing and Clinical Gadgets**

1. **Iron in Clinical Instruments**

1. **Careful Instruments and Iron Compounds**

   Clinical instruments, especially careful apparatuses, as often as possible contain iron combinations. Tempered steel, a well known iron combination in clinical applications, consolidates consumption opposition with strength, making it ideal for careful instruments.

2. Dental Instruments and Iron Parts

Dental instruments, including drills and forceps, frequently consolidate iron parts for their strength and simplicity of disinfection. The dependability of these instruments is significant in keeping up with exclusive requirements of cleanliness in dental methods.

2. Iron in Demonstrative Gear

1. Attractive Reverberation Imaging (X-ray)

One of the momentous uses of iron in medical services is in attractive reverberation imaging (X-ray). The attractive properties of iron assume a significant part in the age of nitty gritty inside pictures, helping with the finding and observing of different ailments.

2. Blood Examination and Iron Reagents

Iron reagents are used in blood examination and research facility diagnostics. These reagents add to the precision and unwavering quality of tests that survey different parts of blood organization, including iron levels.

3. Iron in Prosthetics and Inserts

1. Muscular Embeds and Iron Composites

Muscular inserts, like joint substitutions and bone screws, frequently use iron combinations. The similarity of iron with the human body, combined with its solidarity, makes it a reasonable material for such clinical applications.

2. Hemodialysis and Iron-based Channels

In hemodialysis machines, iron-based channels assume a vital part in eliminating pollutions from the blood. The solidness and biocompatibility of iron add to the viability of these life-saving clinical gadgets.

VII. Difficulties and Contemplations

1. Consumption and Rust Counteraction

1. Defensive Coatings

The vulnerability of iron to consumption, prompting the arrangement of rust, requires the utilization of defensive coatings. Galvanization, painting, and other covering strategies are utilized to protect iron-based objects from natural factors that trigger erosion.

2. Reusing and Manageable Practices

1. Salvaged material Reusing

Advancing the reusing of endlessly iron based materials is fundamental for supportable asset the executives. Salvaged material reusing rations normal assets as well as diminishes the natural effect related with mining and extraction.

3. Mechanical Advancements for Iron Applications

1. High level Iron Amalgams

Progressing research centers around creating progressed iron compounds with upgraded properties. These developments plan to further develop strength, decrease weight, and upgrade erosion obstruction, growing the expected utilizations of iron in different enterprises.

4. Ecological Effect of Iron Extraction

1. Mining Practices and Natural Protection

The extraction of iron metal through mining processes raises worries about ecological corruption. Carrying out reasonable mining rehearses and limiting the environmental impression of iron extraction are basic contemplations for the business.

VIII. Future Viewpoints and Developments

1. Nanotechnology and Iron Nanoparticles

1. Clinical Utilizations of Iron Nanoparticles

Nanotechnology presents the utilization of iron nanoparticles for designated drug conveyance and imaging in the field of medication. These developments hold guarantee for additional compelling and less intrusive clinical medicines.

2. Shrewd Materials and Iron Reconciliation

1. Iron in Shrewd Structure Materials

The combination of iron into brilliant structure materials, equipped for answering ecological circumstances, offers open doors for improved energy proficiency and primary observing in development.

3. Biomedical Advances with Iron

1. Iron-based Therapeutics

Progressions in biomedical examination investigate the advancement of iron-based therapeutics for treating conditions like weakness and lack of iron. These advancements expect to give more designated and effective clinical mediations.

4. Iron in Feasible Energy Stockpiling

1. Iron-particle Batteries

Examination into maintainable energy stockpiling arrangements incorporates the investigation of iron-particle batteries. These batteries can possibly offer an all the more harmless to the ecosystem option in contrast to conventional lithium-particle batteries.

7.2 Magnetic therapy and alternative uses of magnetic fields

Attractive treatment, otherwise called magnet treatment or magnotherapy, is an elective clinical practice that includes the utilization of static attractive fields for restorative purposes. The idea of involving magnets for mending goes back millennia, with verifiable records proposing that antiquated human advancements like the Greeks, Egyptians, and Chinese utilized magnets to address different wellbeing concerns. Lately, attractive treatment has acquired consideration as a reciprocal way to deal with customary medication, and specialists are investigating elective purposes of attractive fields in different fields past wellbeing. This extensive investigation digs into the standards of attractive treatment, its applications in medical services, and the arising elective purposes of attractive fields in assorted areas.

1. Standards of Attractive Treatment

1. Verifiable Foundations of Attractive Treatment

Attractive treatment follows its foundations to old civic establishments where magnets were accepted to have mending properties. The Greeks utilized lodestones, normally happening magnets, for restorative purposes, while Chinese medication consolidated the utilization of attractive stones in needle therapy. In the sixteenth hundred years, Paracelsus, a Swiss doctor and chemist, expounded on the helpful capability of magnets.

2. Attractive Fields and the Human Body

The fundamental reason of attractive treatment is the collaboration between attractive fields and the human body. Promoters of attractive treatment suggest that the body's normal attractive field can be affected by outer magnets, prompting remedial impacts. The attractive fields are remembered to influence the progression of energy, or "chi," in conventional Chinese medication or equilibrium the body's electromagnetic field.

3. Kinds of Magnets Utilized in Attractive Treatment

1. Static Magnets

Static magnets are the most widely recognized type utilized in attractive treatment. These are long-lasting magnets that produce a steady attractive field. They come in different structures, including wristbands, wraps, and sleeping pad cushions, and are frequently applied straightforwardly to the skin over the impacted region.

2. Electromagnetic Treatment

Electromagnetic treatment includes the utilization of electromagnets to make throbbing attractive fields. Dissimilar to static magnets, these gadgets can be acclimated to create various frequencies and forces. Electromagnetic treatment is many times managed utilizing specific machines or gadgets.

II. Uses of Attractive Treatment in Medical services

1. Torment The board

1. Constant Torment Conditions

Attractive treatment is ordinarily utilized for the administration of persistent agony conditions, including joint inflammation, fibromyalgia, and back torment. A few people utilize attractive wraps or sleeping cushions with installed magnets to mitigate inconvenience and advance unwinding.

2. Sports Wounds

Competitors might investigate attractive treatment for the treatment of sports-related wounds. Attractive wraps or fixes are applied to the harmed region with the conviction that the attractive field can upgrade blood flow and diminish irritation.

2. Wound Mending

Attractive treatment has been examined for its likely job in injury mending. A few investigations propose that openness to attractive fields might invigorate cell multiplication and tissue recovery, which could speed up the recuperating system for specific kinds of wounds.

3. Rest Issues

1. A sleeping disorder and Rest Unsettling influences

The quieting impacts of attractive treatment have driven a few people to investigate its utilization for rest issues. Attractive bedding cushions or rest covers with implanted magnets are intended to establish a calming climate and further develop rest quality.

2. Rest related Issues

Conditions, for example, rest apnea have additionally been considered as expected focuses for attractive treatment. Research is progressing to investigate whether attractive excitement can decidedly affect the respiratory muscles and further develop breathing during rest.

4. Neurological Issues

1. **Headaches and Cerebral pains**

   Attractive treatment has been concentrated as an integral methodology for overseeing headaches and cerebral pains. A few people utilize attractive headbands or patches with the expectation of diminishing the recurrence and power of these neurological side effects.

2. **Neuropathic Torment**

Research recommends that attractive treatment might have possible applications in overseeing neuropathic torment, a condition portrayed by harm to the sensory system. Attractive feeling is being investigated as a painless technique to regulate nerve movement.

5. Rheumatic Infections

People with rheumatic infections, like rheumatoid joint pain and osteoarthritis, may go to attractive treatment for help from joint agony and firmness. Attractive arm bands, wraps, or fixes are normally utilized trying to deal with the side effects related with these circumstances.

6. Malignant growth Treatment Backing

Attractive treatment is some of the time investigated as a correlative way to deal with malignant growth treatment. While not an essential therapy for disease, a few people utilize attractive treatment to lighten results of chemotherapy, like queasiness and exhaustion.

III. Arising Elective Purposes of Attractive Fields

1. **Attractive Hyperthermia in Disease Treatment**

1. **Standard of Attractive Hyperthermia**

   Attractive hyperthermia is an arising method that utilizes attractive fields to produce heat inside cancers. Attractive nanoparticles are brought into the cancer, and an outside attractive field is applied to warm the nanoparticles. The raised temperature expects to harm malignant growth cells while saving sound tissue specifically.

2. **Likely Applications and Exploration**

Research in attractive hyperthermia investigates its expected applications in disease treatment. The controlled warming of growths could improve the viability of customary treatments, like radiation and chemotherapy, by expanding their aversion to therapy.

2. Attractive Reverberation directed Centered Ultrasound (MRgFUS)

1. **Joining Attractive Reverberation Imaging (X-ray) and Centered Ultrasound**

   MRgFUS is a painless helpful innovation that joins attractive reverberation imaging (X-ray) with centered ultrasound. Attractive reverberation imaging gives continuous imaging direction, while centered ultrasound creates heat at a particular objective site. This procedure is being researched for different clinical applications.

2. **Clinical Applications**

MRgFUS has been utilized for conditions like uterine fibroids, fundamental quake, and bone metastases. The exact focusing on and harmlessness of this approach make it a promising option in specific clinical mediations.

3. Attractive Excitement in Psychiatry and Nervous system science

1. **Transcranial Attractive Excitement (TMS)**

   Transcranial Attractive Excitement includes the utilization of attractive fields to animate nerve cells in the cerebrum. TMS is a laid out treatment for sadness, and continuous examination investigates its expected applications in other mental and neurological problems.

2. **Neuromodulation**

Attractive fields are likewise being explored for their part in neuromodulation. By affecting brain action, attractive excitement strategies hold guarantee for conditions like epilepsy, Parkinson's sickness, and constant agony.

4. Attractive Field Treatment for Bone Wellbeing

1. **Beat Electromagnetic Field (PEMF) Treatment**

   Beat Electromagnetic Field (PEMF) treatment includes the utilization of electromagnetic fields to animate cell movement. Research recommends that PEMF treatment might have expected applications in bone wellbeing, advancing bone recovery and possibly supporting the treatment of conditions like osteoporosis.

2. **Crack Mending and Osteoarthritis**

Clinical examinations have investigated the utilization of PEMF treatment in speeding up crack mending and overseeing osteoarthritis. The instrument includes the feeling of bone cells and ligament, advancing tissue fix and recovery.

5. Attractive Route in Medication

1. **Attractive Route in Catheterization**

   Attractive route is utilized in operations, especially in catheterization. Attractive fields guide catheters through the vascular framework with accuracy, considering negligibly obtrusive mediations in cardiology and other clinical fortes.

2. **Applications in Endoscopy and Medical procedure**

Past catheterization, attractive route is being investigated in endoscopy and surgeries. The capacity to explore instruments utilizing outer attractive fields offers benefits concerning accuracy and diminished obtrusiveness.

6. Attractive Field Applications in Horticulture

1. **Attractive Field Impacts on Plant Development**

   Research demonstrates that attractive fields can impact plant development and advancement. Attractive field openness has been read up for its capability to upgrade seed germination, increment crop yield, and work on the general strength of plants.

2. **Applications in Soil Remediation**

Attractive fields are additionally being investigated for soil remediation purposes. The utilization of attractive nanoparticles can support the expulsion of impurities from soil, offering an expected harmless to the ecosystem way to deal with address soil contamination.

IV. Contemplations and Difficulties

1. **Logical Examination of Attractive Treatment**

   While attractive treatment has acquired prominence, it stays a subject of logical investigation. Many examinations have created blended results, and established researchers accentuates the requirement for thorough exploration to lay out the adequacy of attractive treatment for different ailments.

2. **Individual Fluctuation Accordingly**

   Reactions to attractive treatment can fluctuate among people. Factors, for example, the strength and length of openness, the particular condition being dealt with, and individual varieties in physiology can impact the results of attractive treatment.

3. **Security Contemplations**

   The security of attractive treatment gadgets is a significant thought. While static magnets utilized in attractive treatment are by and large thought to be ok for transient use, the wellbeing of long haul or extreme focus openness to attractive fields, especially in arising applications, requires cautious assessment.

4. Administrative Status and Normalization

The administrative status of attractive treatment gadgets shifts around the world. Laying out normalized conventions for the utilization of attractive fields in remedial applications is fundamental to guarantee consistency, wellbeing, and adequacy across various gadgets and mediations.

V. Future Headings and Developments

1. **Customized Attractive Treatments**

   Headways in customized medication might prompt the improvement of attractive treatments custom-made to individual hereditary and physiological qualities. Accuracy medication approaches could advance the viability of attractive treatment for explicit people and conditions.

2. **Joining of Attractive Fields in Medical care**

   The joining of attractive fields with other clinical advances, for example, imaging modalities or medication conveyance frameworks, holds guarantee for imaginative medical services arrangements. Joining attractive treatment with laid out clinical practices might improve treatment results.

3. **Investigation of Attractive Fields in Regenerative Medication**

   Scientists are investigating the capability of attractive fields in regenerative medication. Attractive excitement might assume a part in tissue designing, advancing the recovery of harmed tissues and improving the progress of regenerative treatments.

4. **Headways in Attractive Nanoparticle Innovation**

   Headways in attractive nanoparticle innovation might open new roads in medication. Designated drug conveyance frameworks utilizing attractive nanoparticles could upgrade the accuracy and productivity of helpful intercessions.

5. **Attractive Fields in Ecological Remediation**

Further examination into the utilizations of attractive fields in natural remediation could add to supportable answers for soil and water contamination. Attractive innovations might offer productive techniques for the expulsion of poisons from the climate.

7.3 The cultural significance of iron and magnetism

Iron, with its one of a kind properties and attractive charm, has woven itself into the social texture of social orders since forever ago. Past its utilitarian applications, iron and attraction have held representative, legendary, and ceremonial importance across different societies. This investigation dives into the rich woven artwork of the social meaning of iron and attraction, following its strings through folklore, imagery, legends, and strict practices.

1. **Iron in Folklore and Imagery**
1. **Representative Meaning of Iron in Antiquated Societies**
1. **Strength and Perseverance**

   Iron, being one of the most grounded and most sturdy metals, has frequently been emblematically related with strength and perseverance. In antiquated societies, iron apparatuses and weapons addressed the front line of innovation, encapsulating the ability to shape and win.

2. **Profound Cleansing**

The extraordinary course of refining iron mineral into metal mirrors representative subjects of cleaning and change. The demonstration of refining iron was viewed as a figurative sanitization, both in a physical and otherworldly sense, making it an image of profound strength and flexibility.

2. Iron in Fables and Legends

1. **Iron as a Defender**

   Various folktales and legends highlight iron as a focal component, underlining its defensive characteristics. Iron weapons, like swords and horseshoes, are frequently portrayed as instruments that can avoid abhorrent spirits and legendary animals. Iron, in these stories, turns into an image of assurance and security.

2. **Iron as a Slayer of Beasts**

In different social fantasies and legends, iron is depicted as the weapon of decision against powerful elements and beasts. Whether it's a legend employing an iron blade against a mythical serpent or a knight utilizing an iron spear to overcome a mysterious enemy, these accounts build up iron as a power against the obscure and malignant.

II. Iron in Strict Practices

1. **Iron in Ceremonies and Penances**
1. **Contributions of Iron**

   In a few old strict practices, iron assumed a part in contributions and penances. The demonstration of committing iron items to gods was viewed as a token of regard and a method for looking for divine blessing. Iron apparatuses and relics were some of the time remembered for strict services to represent strength and craftsmanship.

2. Consecrated Iron Destinations

Certain areas wealthy in iron stores were viewed as consecrated in different societies. The extraction of iron metal from these destinations was frequently joined by customs and services to mollify the spirits accepted to occupy the land. The relationship of iron with consecrated spaces highlights its importance in strict settings.

2. Iron as an Image of Eternality

1. Iron as a Gift from the Divine beings

In certain folklores, iron was viewed as a gift from the divine beings. Its disclosure and use were frequently credited to help from above or the sharing of information by divinities. This point of view raised iron to a consecrated status, and its usage in strict relics and designs turned into an image of heavenly blessing.

2. Iron in Ceremonial Executes

Strict customs and services frequently elaborate the utilization of iron executes, like ceremonial blades or emblematic articles. These instruments, produced using iron, conveyed profound importance and were accepted to upgrade the adequacy of strict ceremonies. Iron's presence in these customs supported its association with the consecrated.

III. Attraction in Social Convictions

1. Attractive Peculiarities in People Convictions

1. Lodestones and Wizardry

Lodestones, normally polarized bits of iron mineral, have intrigued societies from the beginning of time. In old stories and society convictions, lodestones were frequently connected with enchanted properties. They were accepted to have the capacity to draw in or repulse energy, making them objects of supernatural importance.

2. Navigational Guide in Marine Societies

In marine societies, the attractive properties of lodestones were tackled for route. The utilization of lodestones in compasses permitted mariners to explore the oceans with more noteworthy precision, and the attraction related with these devices became entwined with sea convictions and strange notions.

2. Attractive Imagery in Magical Practices

1. **Attractive Fields as Energy Pathways**

   In magical practices and elusive convictions, attractive fields are at times seen as channels for energy. Practices, for example, energy recuperating or contemplation might include the representation of attractive powers lining up with the body's energy habitats. Attraction, in these specific circumstances, turns into an image of profound arrangement and equilibrium.

2. **Magnets as Charms**

Magnets, with their capacity to draw in and repulse, have been utilized as charms in different social practices. These magnets are accepted to draw positive energy or repulse adverse impacts. The utilization of magnets as defensive charms mirrors the social faith in the impact of attractive powers on prosperity.

IV. Iron and Attraction in Workmanship and Style

1. **Iron as Imaginative Medium**

1. **Iron Figures and Relics**

   Iron's flexibility and strength make it a flexible mechanism for creative articulation. Iron models, fashioned ironwork, and imaginative executes exhibit the craftsmanship and stylish allure of iron. In social settings, iron imaginativeness reflects both utilitarian and representative aspects.

2. **Iron in Engineering and Plan**

Compositional wonders frequently consolidate iron components, from luxurious iron doors to primary parts. Iron's solidness and capacity to be formed into many-sided plans make it an inclined toward material in the production of notorious designs. The social meaning of iron is implanted in the style of these engineering works of art.

2. Attractive Craftsmanship Establishments

1. **Attraction as a Creative Idea**

   Contemporary specialists investigate the idea of attraction as a subject in their works. Establishments including attractive fields, intelligent attractive figures, and works of art impacted by attractive peculiarities give a tactile encounter that obscures the lines among science and craftsmanship. These manifestations welcome watchers to draw in with the imperceptible powers of attraction in an instinctive and creative manner.

2. **Intelligent Attractive Displays**

Attractive displays in exhibition halls and workmanship displays offer intuitive encounters for guests. Active presentations including attractive fields, suspending objects, or attractive liquid models engage as well as instruct about the standards of attraction. These displays overcome any barrier among science and culture, encouraging a more profound appreciation for attractive peculiarities.

V. Social Customs and Customs

1. Iron in Transitional experiences

1. Iron in Birth Customs

In specific societies, iron is integrated into customs related with labor. Iron items or images might be utilized to favor and safeguard babies, representing the expectation for an existence of solidarity and strength.

2. Iron in Transitioning Functions

Transitioning functions in different societies might include the giving of iron articles or the emblematic passing down of iron devices. These ceremonies mean the change to adulthood and the procurement of solidarity and obligation.

2. Attractive Customs for Recuperating

1. Attractive Mending Practices

A few social practices integrate attractive customs for the end goal of mending. Professionals might involve magnets or attractive fields as a feature of all encompassing recuperating services, trusting in the restorative properties of attraction for physical and otherworldly prosperity.

2. Polarized Articles in Ceremonies

Objects with attractive properties, like lodestones, may assume a part in ceremonies pointed toward drawing in sure energy or dispersing pessimism. The ceremonial utilization of polarized objects reflects social convictions in the impact of attractive powers on the otherworldly domain.

www.ingramcontent.com/pod-product-compliance
Lightning Source LLC
Chambersburg PA
CBHW071159130726
47998CB00002B/548

AI IN MIND MAPPING